Integration of Robots into CIM

Integration of Robots into CIM

Edited by
R. Bernhardt
Researcher IPK Berlin, Germany

R. Dillman
Professor in Computer Science Faculty for Informatics
Institute for Real-Time Computer Control Systems and Robotics,
University of Karlsruhe, Germany

K. Hörmann
Assistant Professor of Computer Science Faculty for Informatics
Institute for Real-Time Computer Control Systems and Robotics
University of Karlsruhe, Germany

K. Tierney
Manager of CIM Research Unit University College
Galway, Ireland

CHAPMAN & HALL
London · New York · Tokyo · Melbourne · Madras

Published by Chapman & Hall, 2–6 Boundary Row, London SE1 8HN

Chapman & Hall, 2–6 Boundary Row, London SE1 8HN, UK

Van Nostrand Reinhold Inc., 115 5th Avenue, New York NY10003, USA

Chapman & Hall Japan, Thomson Publishing Japan, Hirakawacho Nemoto Building, 7F, 1-7-11 Hirakawa-cho, Chiyoda-ku, Tokyo 102, Japan

Chapman & Hall Australia, Thomas Nelson Australia, 102 Dodds Street, South Melbourne, Victoria 3205, Australia

Chapman & Hall India, R. Seshadri, 32 Second Main Road, CIT East, Madras 600 035, India

First edition 1992

© 1992 Chapman & Hall

Printed in Great Britain by T.J. Press (Padstow) Ltd, Padstow, Cornwall.

ISBN 0 412 37140 5   0 442 31243 5 (USA)

A catalogue record for this book is available from the British Library

Library of Congress Cataloging-in-Publication data
Integration of robots into CIM / edited by R. Bernhardt . . . [et al.].
    — 1st ed.
        p. cm.
    Includes bibliographical references and index.
    ISBN 0-442-31243-1 (alk. paper)
    1. Computer integrated manufacturing systems. 2. Robots, Industrial. I. Bernhardt, Rolf. 1934–
TS155.6.I5535   1992
670.42'72—dc20   91–30481
                                                          CIP

Contents

Contributors

Dr. Rolf Bernhardt
IPK
Fraunhofer-Institut für Produktionsaniagenund Konstruktionstechnik
Pascalstraße 8–9
D – 1000 Berlin 10, Germany

Mr Paolo Bison
LADSEB
Consiglio Nazionale delle Ricerche
Istituto per Ricerche di Dinamica del Sistemi
Corso Stati Uniti 4
1 – 35020 Padova, Italy

Prof. Dr. Jim Browne
UCG
University College Galway
Dep. of Industrial Engineering
CIM Research Unit
Nun's Island
Galway
Eire

Mr V. Caglioti
POLIMI
Politechnico di Milano
Dip. di Elettronica
Piazza L. Da Vinci 32
1 – 20 133 Milano
Italy

Dr Luis M. Camarinha-Matos
PSI
Gesellschaft für Prozeßsteuerungsund Informationssyteme mbH
AT-2
Heilbronner Straße 10
D – 1000 Berlin 31
Germany

Prof. Dr. Rüdiger Dillman
University of Karlsruhe
Institute for Process Computer
Control and Robotics
Postfach 6980
D 7580, Karlsruhe 1, Germany

Mr Bertrand Duffau
Renault Automation
Sa 0981
8 – 10 avenue Emile-Zola
F – 92109 Boulogne Billancourt
Cedex, France

Mr B. Frommherz
UKA
University of Karlsruhe
Institute for Process Computer
Control and Robotics
Postfach 6980
D 7580 Karlsruhe 1,
Germany

Mr R. Gallerini
FIAR
Via Mercantesse 3
1 – 20021 Baranzate
Milano, Italy

Mr Volker Gleue
IPK Fraunhofer-Institut für Produktionsaniagen und Konstruktionstechnik
Pascalstraße 8–9
D – 1000 Berlin 10
Germany

Dr Klaus Hörmann
Forschunzenstrum Informatik
Hald-u. -Neu-Str. 10–14
D – 7500 Karlsruhe 1
Germany

Mr J. Hornberger
UKA
University of Karlsruhe
Institute for Process Computer
Control and Robotics
Postfach 6980
D 7580 Karlsruhe 1,
Germany

Mr W. Jakob
PSI
Gesellschaft für Prozeßsteuerungsund Informationssyteme mbH
AT-2
Heilbronner Straße 10
D – 1000 Berlin 31
Germany

Prof. Dr. A. Jimenez
UPM
Universidad Politecnica de Madrid
Departamento de Ingenieria de Sistemas y Automatica
C./J. Gutierrez Abascal 2
E – 28 006 Madrid
Spain

Mr Volker Katschinski
IPK
Fraunhofer-Institut für Produktionsaniagen und Konstruktionstechnik
Pascalstraße 8–9
D – 1000 Berlin 10
Germany

Mr B. Kottke
PSI
Gesellschaft für Prozeßsteurungsund Informationssysteme mbH
AT-2
Heilbronner Straße 10
D – 1000 Berlin 31
Germany

Mr Stephan Krüger
IPK
Fraunhofer-Institut für Produktionsaniagen und Konstruktionstechnik
Pascalstraße 8–9
D-1000 Berlin 10
Germany

Mr Geleyn R. Meijer
UA
University of Amsterdam
FWI
Kruislaan 403
NL – 1098 SJ Amsterdam
Netherlands

Mr C. Mirolo
LADSEB
Consiglio Nazionale delle Ricerche
Istituto per Ricerche di Dinamica del Sistemi
Corso Stati Uniti 4
1-35020 Padova
Italy

Prof. Dr. E. Pagello
LADSEB
Consiglio Nazionale delle Ricerche
Istituto per Ricerche di Dinamica del Sistemi
Corso Stati Uniti 4
1-35020 Padova
Italy

Mr A. Pezzinga
FIAR
Via Mercantesse 3
1 – 20021 Baranzate
Milano
Italy

Prof. Marco Somaivico
POLIMI
Politecnico di Milano
Dip. di Elettronica
Piazza L. Da Vinci 32
1 - 20 133 Milano
Italy

Ms L. Stocchiero
LADSEB
Consiglio Nazionale delle Ricerche
Istituto per Ricerche di Dinamica del Sistem
Corso Stati Uniti 4
1 - 35020
Padova
Italy

Ms K. Tierney
UCG
University College Galway
Dep. of Industrial Engineering
CIM Research Unit
Nun's Island
Galway
Eire

Mr Ugo Negretto
UKA
University of Karlsruhe
Institute for Process computer
Control and Robotics
Postfach 6980
D 7580
Karlsruhe 1
Germany

Mr S. Wadwha
UCG
University College Galway
Dep. of Industrial Engineering
CIM Research Unit
Nun's Island Galway
Eire

Dr Francisco Sastrón
UPM
Universidad Politecnica de Madrid
Departamento de Ingenieria de Sistemas y Automatica
C./J. Guiterrez Abascal 2
E – 28 006 Madrid
Spain

Mr Gerhard Schrek
IPK
Fraunhofer-Institut für Produktionsaniagen und Konstruktionstechnik
Pascalstraße 8–9
D – 1000 Berlin 10
Germany

Mr K. Schröer
IPK
Fraunhofer-Institut für Produktionsaniagen und Konstruktionstechnik
Pascalstraße 8–9
D – 1000 Berlin 10
Germany

Mr Georg Stark
KUKA
Schweißaniagen & Roboter GmbH
Abt. STK-EP
Blücherstraße 144
D – 8900 Augsburg
Germany

Prof. Dr. A. Steiger-Garçáo
UNL
Universidade Nova de Lisboa
FCT – Departamento de Informatica
Quinta da Torre
P-2825 Monte Caparica
Portugal

Mr Peter Welgele
KUKA
Schweißanlagen & Roboter GmbH
Abt. STK-EP
Blücherstraße 144
D-8900 Augsburg
Germany

Mr G. Werling
UKA
University of Karlsruhe
Institute for Process Computer
Control and Robotics
Postfach 6980
D 7580 Karlsruhe 1
Germany

FOREWORD

From its inception in 1983, ESPRIT (the European Strategic Programme for Research and Development in Information Technology) has aimed at improving the competitiveness of European industry and providing it with the technology needed for the 1990s.

Esprit Project 623, on which most of the work presented in this book is based, was one of the key projects in the ESPRIT area, Computer Integrated Manufacturing (CIM). From its beginnings in 1985, it brought together a team of researchers from industry, research institutes and universities to explore and develop a critical stream of advanced manufacturing technology that would be timely and mature for industrial exploitation in a five year time frame. The synergy of cross border collaboration between technology users and vendors has led to results ranging from new and improved products to training courses given at universities.

The subject of Esprit Project 623 was the integration of robots into manufacturing environments. Robots are a vital element in flexible automation and can contribute substantially to manufacturing efficiency. The project had two main themes, off-line programming and robot system planning. Off-line programming enlarges the application area of robots and opens up new possibilities in domains such as laser cutting, and other hazardous operations. Reported benefits obtained from off-line programming include:

- significant cost reductions because re-programming eliminates robot down-time;
- faster production cycles, in some cases time-savings of up to 85% are reported;
- the optimal engineering of products with improved quality.

Moreover, off-line programming techniques protect the operator, who under conventional systems of on-line programming, is at a risk of injury from having to work in the physical proximity of the robot.

The integration of robots in manufacturing cells requires the integration of information concerning product design, plant availability and system layout. Project 623 has achieved this through the use of relational and knowledge databases which lead to large cost savings for vendors providing turnkey systems and users who need fast adaptation to production demands.

The project has been an excellent example of a multi-disciplinary approach, combining the knowhow of mechanical and manufacturing engineers and computer scientists to push forward the frontiers of knowledge in an application domain which is at the leading edge of the major European economies.

Section I

Survey of the project

Chapter 1

Robots in CIM

R. Bernhardt
IPK Berlin, Germany

Information technology has initiated a structural change in manufacturing industry. Productivity, flexibility, quality, and reliability can now attain a level which could not be realized on the basis of conventional production structures. The very different technological requirements of products has led to increased product variety and shorter product life cycles. These alterations of the market situation demand automation of the highest flexibility and productivity. This can be reached by computer integrated, automated and flexible manufacturing, whereby information techniques take over a key function.

In production technology it was comprehended at a very early stage that computers are important components. Stages of these developments were the NC technique, CNC controls and the development of FMS with robots. In this context it can be stated that robots play an important role as the most flexible automation components. This is also documented by the growing number of robot applications as shown in Fig. 1.1. Specifically remarkable is thereby the big difference between the USA and Europe on the one hand and Japan on the other hand which shows impressively the development potential of the market.

The growth rate shown in Fig. 1.1 has been reached mainly by applying state-of-the-art automation techniques, i.e. without the consideration of the aspect 'robot as CIM component'. But this means that the immanent flexibility of robot systems has been used only to a marginal extent. To use effectively this basic feature, on the one hand powerful, computer-aided tools for planning, programming and simulation must be available. On the other hand, robot controllers must be much improved and enlarged with regard to functionality, open system aspects, user support, communication features and the integration of sensors.

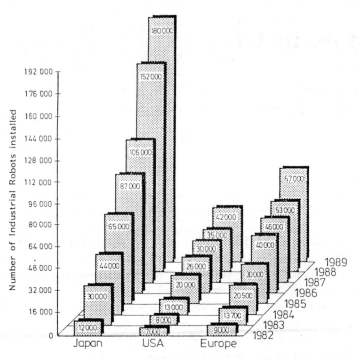

Fig. 1.1   International distribution of robots (IPA Stuttgart).

The first area mentioned above does not only mean that those tools are available to support planning engineers, but also includes that these tools are integrated into an information system which guarantees a continuous information flow from design to the shop floor. This is sometimes called 'information integration' and is probably the most important aspect covered by the term CIM.

The development of such computer-aided tools, their integration and the demonstration of the systems under realistic industrial conditions formed the main objectives of the European research and development project, entitled *Operational Control of Robot System Integration Into CIM*. It was was conducted in the frame of the ESPRIT programme. This book reports on the work done in the project. As can be concluded from the project title, a strong emphasis was put on the component robot. From a general automation point of view, robots are important but still only one among other CIM components which are required for advanced production systems. This means that for efficient automation planning engineers have to be supported by computer-aided tools for the whole of their tasks of

4

planning, programming, testing, installing, operating and maintaining production systems. Furthermore, they must be enabled to re-plan an existing and running manufacturing system at any of the levels mentioned above. Concerning these topics, another R&D project has been launched in the meantime, also within the ESPRIT framework. The title of this project is CIM System Planning Toolbox (CIM-PLATO, ESPRIT Proj. No. 2202).

Both projects turned out to suffer very much from the fact that computer simulation models are not commercially available. This concerns e.g. models of robot controllers, kinematics, etc. To make those models available for the market, a step which substantially enlarges the product range of robot and/or controller manufacturers, strong efforts in the standardization area have to be undertaken. This also formed a work area in the project reported on here. Furthermore, another project has been launched (CIMDATA, ESPRIT) with the objective of realizing a database containing CIM components. The overall idea behind it is to provide planning engineers with all the necessary computer simulation models on the various planning levels which they need to fulfil their entire production system planning task.

Finally it may be mentioned that robots play an important role for automation, but in order to become a real CIM component, they still have to be improved. During the last years, many efforts have been undertaken for the realization of tools to support the planning of robotized systems. This work has to be taken further and enlarged with regard to items such as automated, optimal and intelligent planning functions and procedures.

But for the system robot itself, a lot of R&D work has to be done, too, especially for the development of advanced robot controllers according to the previously mentioned topics which may be summarized under the term 'open system architecture'.

The availability of powerful computer-aided tools which are integrated in a CIM environment ease the task of planning and programming robotized systems. These integrated tools represent advanced automation techniques. But on the other hand the 'real' components such as robot controllers have to be on the same advanced level. Only when this is reached, the system robot will be a true CIM component. The integration of robots into CIM systems as the subject of this book, is based on the R&D work done in the ESPRIT project No. 623. In this first section, the objectives, approaches and benefits gained within this project are described, the project partners are presented, and the overall structure of the book is explained.

5

Chapter 2

Objectives, approaches and main benefits

R. Bernhardt
IPK Berlin, Germany

2.1 Introduction

Robots are important components for flexible automation. The enlargement of their application as well as their integration into a CIM environment requires the availability of computer aided means for planning and programming of robotized manufacturing systems. To develop these means, an R&D project in the frame of the European Strategic Program of Research and Development in Information Technology (ESPRIT) was started in 1985 [1,2]. This project entitled

'Operational Control for Robot System Integration into CIM, Systems Planning, Implicit and Explicit Programming'

was finished in 1990. Its general goal was to implement industrial prototypes which demonstrate the integration of robot systems into a CIM environment [3]. The critical path of this integration concerns the operational level of CIM, and includes two closely interrelated fields of R&D: the implementation of a computer aided planning system for robotized work cells and of an off-line programming system for robots (Figure 2.1). Both fields were identified and analyzed in detail in the frame of the ESPRIT project 75:

'Design Rules for the Integration of Industrial Robots into CIM Systems'
in which also the main objectives of the project described in this book were determined.

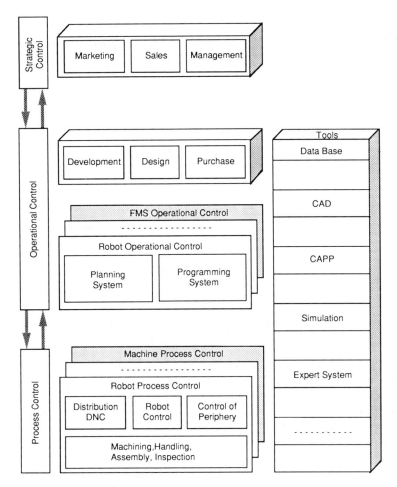

Fig. 2.1   Project reference model.

Based on this reference model, the project objectives are outlined in more detail in the subsequent paragraphs.

2.2   Objectives and approaches

Based on the general goal to implement computer aided planning and off-line programming systems, more detailed objectives were identified.

7

Planning was understood to be an integrated procedure starting with general project information, requirements list, and work piece data and resulting in optimized layout data of a robotized work cell as well as a formalized description of the task to be executed within this work cell. To explain this in more detail the case of planning robotized assembly cells has been chosen.

The integrated procedure starts with the generation of an assembly sequence plan which includes operation sequence planning, precedence analysis, preselection of robots which are suitable for the different assembly steps and selection of available feeders and grippers. Based on the assembly sequence and the suggested devices, components are selected and roughly arranged. In an interactive and iterative procedure this layout is improved and verified using supporting tools (e.g. for cycle time estimation and material flow simulation). In the next step the generated planning data are transformed into a structure suitable for the off-line programming procedure. In addition to the layout description, the assembly sequence is described by a semiformal representation containing also geometrical, technological and control specific data. The planning phase is completed with a layout optimization procedure considering type variants and locations of robots, tools and peripheral devices, and their influence on important parameters ranging from cycle times to the total cost of the work cell.

The general objective for the development of a planning system was the realization of computer-aided tools to support the procedure explained above. In this context, off-line programming means the detailed planning of the tasks to be executed by the robot(s) before the real work cell is built on the shop floor. (Sometimes in literature the term *execution planning* is used instead of *programming*). This includes the planning of the geometrical path, the determination of motion and technology parameters, the control of interactions with peripherals and the generation of the robot program code (application program). The correctness and executability of the generated application programs have to be tested before they can be transferred to the real robot. Therefore, a process execution simulation has to be an integral part of the programming system.

For this purpose, models and algorithms have to be available which are computer internal representations describing all work cell components with regard to their motion behavior (control models), kinematics, and shapes. In addition to these basic functions of an off-line programming system for robots, three more objectives were identified: trajectory optimization with an eye on cost criteria, minimal stress of joints and/or minimal cycle time, and handling of exceptional cases occurring in the work cell by *a priori* planned procedures.

8

As far as the concrete work in the project was concerned, it was decided to follow two different approaches for the implementation of off-line programming systems.

The first approach was aimed at realizing an implicit programming system which allows the determination of the robot's task by a task level instead of a robot level specification. Hence the task is described as a sequence of changes in a model of the environment. An action is characterized by its desired effects on objects of the model rather than by its specified motions in order to achieve a task The system then plans these actions and generates a robot level representation of the task [4, 5].

The second approach concerned the development of an explicit programming system based on an interactive procedure. This means that all the necessary actions have to be specified explicitly by the user. A verification of each step in this interactive procedure is enabled by the simulation and test system through visualization capabilities. This allows, for example, a robot independent test of correct task sequences or the visualization of the robots in motion in the production environment [6]. Again, the output of this system is a robot level representation of the task.

One important goal of the project was to specify the task representation at the robot level described above, independent of a specific robot language. This data structure is called *explicit solution representation (ESR)* and contains, at the end of the programming procedure (either explicit or implicit), all necessary information to generate an application program for a specific robot controller.

A further objective was to allow the exchange of modules developed either for the implicit or for the explicit programming system. As a result, robot code generators realized in the explicit area can be used within the implicit system. Another example is the use of implicit modules for trajectory optimization which upgrade the functionality of the explicit programming system. This could be achieved by modular system structures and clearly specified interfaces. In this sense, both approaches are aimed in the same direction. The first one is a top-down approach starting with a specification of the entire implicit off-line programming system in general. However, the realization is adapted and limited to what is practicable today. In this respect it is a more research oriented approach. The second approach is characterized by the practical constraints put forth by industry. It takes into account the limitations during the specifications of modules, but also can interface with the integration of implicit functions. Thus this bottom-up approach in the end will also lead to an implicit programming system by upgrading the explicit programming system through the replacement of interactive functions by automatic ones.

To reach the objective of realizing a flexible system structure which facilitates quick adaption of the system to user needs, a specific working group 'information integration' was installed. These aspects were studied in great detail especially in the second half of the project.

As a summary of the objectives of the first half of the project, the following prototypes were planned:

- a planning system for robotized work cells;
- an explicit or motion oriented programming and simulation system;
- an implicit or task oriented programming system.

Work in the three areas proceeded simultaneously. In all cases, definition and specification of software modules was followed by their realization and application/demonstration in real or simulated industrial environments.

The objective for the second half of the project was the creation of industrial and research prototypes of an integrated planning and programming system for robotized cells. This encompassed on the one hand, the integration of the subsystems mentioned above and on the other hand, the integration of the whole system into a CIM environment.

The first aspect concerns mainly the detailed definition of information interfaces between the planning and the programming system. In Figure 2.2, the main information interfaces of the different subsystems are shown by using the form of an ISAC diagram, in which a block 'translation, robot interface' is added since it was also an objective of the project to generate executable application programs and to transfer them to the real robot.

For simplification, all information needed for the activities characterized by rectangular blocks are marked as 'auxiliary information'. This is not explained in detail, as the diagram should only stress the information interfaces between subsystems. The most important information produced by the planning system are the formal descriptions of the workcell layout and the production task to be executed. By using this information, an explicit solution representation can be produced. The ESR is a system-internal information representation which contains at the end of the programming procedure (either explicit or implicit) all information required for the automatic generation of application programs for robots. Before these programs are transferred to the real robot, they are tested by a simulation system which is an integral part of the programming system. To what extent the generated application programs can be executed in the real work cell depends on the work cell information available during planning and programming. This concerns the absolute positioning accuracy of the robot itself but also the other components and their arrangement in the work cell.

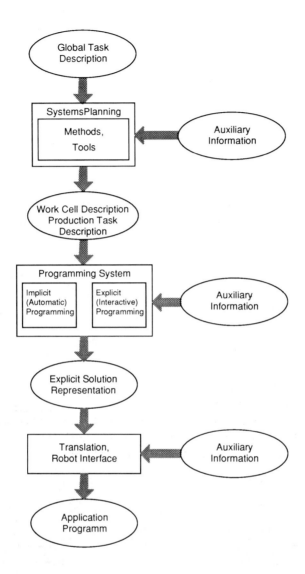

Fig. 2.2   Main information interfaces between subsystems.

The second aspect concerns the information integration into a CIM environment, i.e. the integration into a company wide (or even wider) information system. This is illustrated in a simplified manner in Figure 2.3.

11

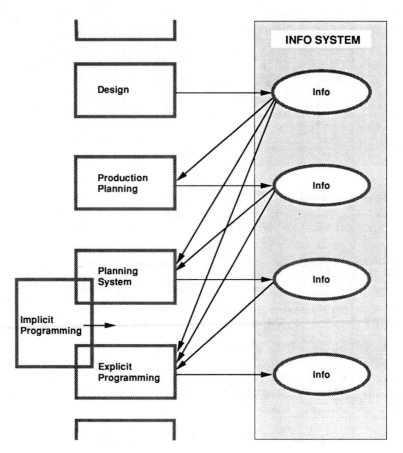

Fig. 2.3   Information integration in a CIM environment.

Again, the rectangular blocks representing activities produce information which is consumed by one or more other activities. It should be stressed that a direct information exchange between subsystems was not envisaged but rather the exchange via a central information system. This was introduced for information management and consistency purposes. For the realization of industrial prototypes, a relational data base management system (DBMS) was used.

For more research oriented prototypes, an information system was implemented combining the technologies of knowledge based systems (KBS) and DBMS. The so-called knowledge and information system (KIS), while also a kernel for the integration of all functional modules, primarily aims at a higher degree of flexibility.

The following example explains the ideas represented in Figure 2.3. Data available in the design area (CAD system) and data from the production planning area (technology data, production facility data) can be accessed by the planning system via the information system. Within the DBMS, the user rights are also defined (e.g. to read, write or alter data). Therefore not only an information exchange is required, but also a control flow (feedback), such as for inquiring during planning in the design area about possibilities to re-design a workpiece which would lead to a less complex layout.

To keep Figure 2.3 more transparent the feedback loops are not shown. Finally, the block 'implicit programming' should be mentioned. The overlapping between planning and explicit programming should express, on the one hand, that there are some functions of implicit programming which can also be located in the other blocks, and on the other hand, it should be stressed that planning and programming are to be seen as an integrated procedure.

For the realization of these basic objectives to integrate subsystems, and for the integration of information into a CIM environment, demonstrator concepts were elaborated.

A further objective was to show the increased functionality and efficiency of an integrated planning and programming procedure for robotized work cells. This also included demonstrating the integratability of modules and subsystems realized by different project partners.

The objectives for the second half of the project clearly show that the information integration aspects were of major importance for the success of the project. This concerns the information and control flow within the integrated systems, the integration into a CIM environment, and also standardization aspects. Another objective was the development of advanced information system concepts (knowledge based approaches) and their implementation.

For the realization of the demonstrators, different principles like automatic or interactive planning functions and explicit or implicit programming procedures had to be considered. Additional realistic industrial applications or well-known benchmark tests had to be chosen. A further important criterion for selecting applications was to cover a broad spectrum of problems and show their solutions. Therefore applications were selected with:

- Parts of very different size requiring gripper exchanges.
- Parts of different shapes requiring multifunctional grippers.
- Assembly operations of different mating directions which require complex fixtures and robots with six degrees of freedom.

13

For the realization of demonstrators, the different interests of industrial and academic partners also had to be taken into account. It was intended that the demonstrators be useful for the industrial partners from different branches, and at the same time that they form a suitable basis for future research from the academic partners' point of view. To consider these various interests, it was decided to create four demonstrator systems, each of which was to highlight specific aspects, such as information exchange between different partners at different locations (distributed system), realization of advanced implicit programming, or direct integration into the everyday work in a factory.

2.3 Main results

As the partners who worked together in the project were not only from industry but also from academia, it was necessary to differentiate between results and benefits seen mainly from the industrial point of view and those of more concern to research institutions and universities [8,9].

First, the results and benefits of the industrial partners within the project were presented. Computer aided tools were developed to support the various planning and programming activities in the companies. These prototype products resulting from the project work were customised and extended by each partner. They then were introduced into the planning and programming departments and applied to many industrial projects. The result was higher productivity on the suppliers' side and better quality of the delivered manufacturing systems.

Two kinds of products were developed in order to serve the planning and programming needs. In the first category, there are integrated planning and off-line programming systems using simulation models which are applied in the office and run on work stations or minicomputers. In the second are application support systems that run on small, portable computers. They are used for adapting systems on the shop floor.

To support planning tasks, robot simulation is the most important method. The most frequent kinds of applications are:

- Tasks with critical time specifications.
- Complex robot operations with multi-purpose endeffectors.
- Complex kinematic structures with more than six joints involved.

Also for off-line programming, industrial applications were carried out. Three categories were identified where the use of these techniques is most advantageous for the industry:

14

- Programming of regular (e.g. symmetric) workpieces by applying program generation or manipulation functions (e.g. mirror function).
- High accuracy applications (e.g. laser beam cutting) requiring joint level program optimization.
- Transfer and adaptation of existing programs to similar applications (e.g. adaptation of a transfer line to a new car model).

To evaluate the results and benefits from an industrial point of view, the most important qualitative advantages are the following:

- Personnel costs for design, optimization and programming are saved and costs for workshop tests and subsequent alterations of the systems are significantly reduced.
- At a very early stage in a project, highly accurate information is available (e.g. with regard to the configuration of the layout).
- User solutions are developed faster, also within the tender preparation phase.
- Several alternative solutions can be analyzed and compared, resulting in a decisive improvement of the quality of a solution.

To quantify the results and benefits, the industrial partners carried out inquiries. To give an example, some results of an inquiry (end of 1989) by the project partner KUKA are presented in Figure 2.4. As can be seen, time savings up to 90% could be reached by using the systems realized in the course of the project.

In close cooperation with the industrial partners two approaches were followed by the academic partners during the course of the project. In the first approach, existing techniques were used for test and integration into prototype systems. These systems were installed and evaluated together with the industrial partners. The second approach aimed at developing new technologies and showing their usefulness in experiments. In this way, a proper input for the development of prototypes was ensured.

The objective of the research and the development activities was to demonstrate the applicability of advanced technologies in the areas of planning and programming for operational robot control. Thereby two major areas were addressed. The first concerned the development of an automatic planning and programming system and the gradual replacement of interactive systems by automatic ones. Secondly, information integration aspects were studied in detail. In particular in the later phases of the project, the research efforts were directed at this second area. Throughout the project close contact with other research groups outside the consortium was maintained.

15

Type of Application

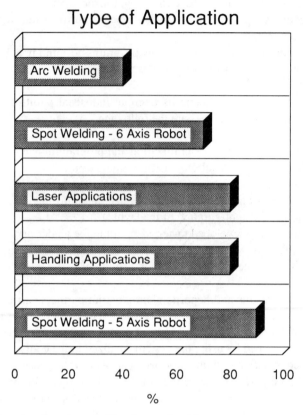

Fig. 2.4 Time saving by application robot programming and simulation (KUKA).

In total, 30 products and services resulted from the academic partners. From these, 20 products and services were delivered to external institutions and companies. In particular the research experiences were used for other application areas such as automation of robotics in space and underwater applications. The academic partners were able to start 14 new industrial cooperations and 26 public projects in national and European programs.

An important aspect for the academic partners was the dissemination of knowledge through courses, publications and conferences. New courses were set up in the areas of CIM systems, robot programming and also in new subjects of sensors and artificial intelligence (Figure 2.5).

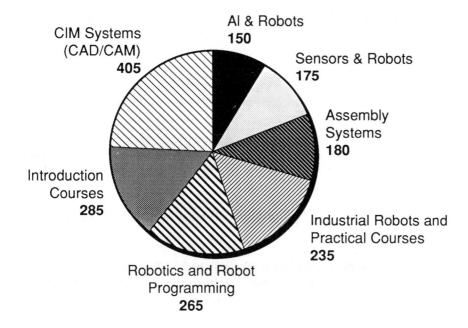

CIM Systems
(CAD/CAM)
405

AI & Robots
150

Sensors & Robots
175

Assembly
Systems
180

Introduction
Courses
285

Industrial Robots and
Practical Courses
235

Robotics and Robot
Programming
265

Fig. 2.5   Student qualification (1989).

Over 100 publications were edited by the partners and more than 50 papers were presented at international conferences. In particular during the last period the consortium organized a number of workshops and seminars and presented the achieved results. In the course of the project topical cooperations between academic and industrial partners have been established which exceeded the scope of the project. In the future, these research cooperations will also be used to address the new challenges in robot applications.

The basic research efforts of the academic partners are now focussing on the new emerging technologies. These include the application of expert system techniques and advanced programming approaches like object oriented formalisms.

References

[1] Spur, G. *et al.*, Integration of Industrial Robots into CIM-Systems. SITEF Toulouse, Oct. 24th, 1986.

[2] Rembold, U., Vojnovic, M., *Operational Control for Robot System Integration into CIM*. IEEE 1986, Interantional Conf. on Robotics and Automation, April 7-10, 1986, San Francisco.

[3] Spur, G. *et al.*, Planning and Programming of Robot Integrated Production Cells. Proceedings of the 4th Annual ESPRIT Conference, Brussels, September 28-29, 1987, North-Holland.

[4] Frommherz, B., Hoermann, K., *Ein Konzept für ein Roboter. Aktionsplanungssystem.* 16. Jahrestagung der Gesellschaft für Informatik, Berlin, 1986.

[5] Brady, M. *et al.*, Robot Motion: Planning and Control. The MIT Press, 1982.

[6] Bernhardt, R., *Integrated Planning and off-line Programming System for Robotized Work Cells*. Proceedings of the 6th Anual ESPRIT Conference, Nov 27-Dec 1, 1989, Brussels.

[7] Duelen, G., Schröer, K., *Robot Calibration - Method and Results.* Robotics and Computer-Integrated Manufacturing 8, 1991.

[8] Operational Control for Robot Systems into CIM, Results and Benefits, 10th Interim Report of the ESPRIT 623 Project, Nov. 1989.

[9] Robot System Integration into CIM, Workshop Proceedings, March 1990, Berlin.

Chapter 3

Introduction and role of the partners

R. Bernhardt
IPK Berlin, Germany

The ESPRIT 623 project involved twelve partners from seven European countries: eight partners from research institutions and universities and four partners from the industry. In Table 3.1 a list of all partners is presented.

The project was executed in two main phases. In the first phase the three subsystems, planning, implicit, and explicit programming were realized. To achieve this, a subgroup structure was installed which is outlined in Table 3.2.

In the second phase subsystems and components had to be connected and integrated in a CIM environment. Therefore demonstrator projects were established (Table 3.3).

The cooperation and information exchange between the partners was organized on three levels: meetings, exchange of papers, and exchange of software.

The following meetings were held periodically:

- General meetings for the discussion of topics relevant for the whole project; these meetings were chaired by the project manager.
- Subgroup/demonstrator group meetings for the discussion of technical topics relevant for one group: these meetings were chaired by the responsible subgroup speaker/demonstrator project manager.

The results of these meetings were distributed in the form of minutes to all project participants in accordance with an agreed procedure. Besides these meetings bi- or multilateral meetings were held when required.

Additionally to meetings, working papers were prepared and circulated to keep other partners informed about the ongoing work.

Furthermore every half year a project report was prepared containing contributions of all partners.

For the realization of software, specifications were elaborated jointly and for integration purposes software modules were exchanged.

In Table 3.4, a project organigram is presented which shows on the one hand the internal project steering mechanisms and on the other hand the project control by external reviewers and the project officer from the CEC.

In the course of the project an external review took place every half year at which project results were presented and demonstrated.

Prime Contractor:	-	Institut für Produktionsanlagen und Konstruktions-technik (IPK), Berlin, Federal Republic of Germany
Contractors:	-	KUKA Schweißanlagen und Robotertechnik GmbH, Augsburg, Federal Republic of Germany
	-	Renault Automation, Paris, France (RA)
	-	University College Galway, Republic of Ireland (UCG)
	-	FIAR - SPA, Divisione Robotica Industriale, Milano, Italy
	-	Lehrstuhl für Prozeßrechentechnik, Universität Karlsruhe, Federal Republic of Germany (UKA)
	-	Universidad Politecnica de Madrid, Spain (UPM)
	-	Universidade Nova de Lisboa, Portugal (UNL)
Subcontractors:	-	PSI - Gesellschaft für Prozeßsteuerung und Informationssysteme mbH, Berlin, Federal Republic of Germany
	-	Politecnico di Milano, Italy
	-	LADSEB - CNR, Padova, Italy (PM)
	-	Universiteit van Amsterdam, Netherlands (UA)

Table 3.1:   Project partners and status.

Systems-Planning

<u>Speaker</u>:   Mrs Tierney, UCG

<u>Participants</u>:   - KUKA Schweißanlagen und Robotertechnik GmbH
                  - University College Galway (UCG)
                  - Renault Automation (RA)
                  - Institut für Produktionsanlagen und Konstruktions-technik (IPK)
                  - Universidad Politecnica de Madrid (UPM)

Explicit Programming

<u>Speaker</u>:   Prof. Dillmann, UKA

<u>Participants</u>:   - Universität Karlsruhe (UKA)
                  - Renault Automation
                  - Institut für Produktionsanlagen und Konstruktionstechnik (IPK)
                     * Universiteit van Amsterdam (UA) as subcontractor of IPK
                  - Universidade Nova de Lisboa (UNL)

Implicit Programming

<u>Speaker</u>:   Dr. Hörmann, UKA

<u>Participants</u>:   - Universität Karlsruhe (UKA)
                     * PSI - Gesellschaft für Prozeßsteuerung und Informationssysteme mbH as subcontractor of UKA
                  - FIAR - SPA, Divisione Robotica Industriale
                     * Politecnico di Milano (PM) as subcontractors of FIAR
                     * LADSEB - CN as subcontractors of FIAR
                  - Institut für Produktionsanlagen und Konstruktionstechnik (IPK)

Table 3.2:   Subground structure.

Demonstrator A: Integrated Planning and Explicit Off-Line Programming System

Demonstrator
Project Manager: Mr Stark, KUKA

Participants:
- KUKA Schweißanlagen und Robotertechnik GmbH
- Gesellschaft für Prozeßsteuerung und Informationssysteme mbH (PSI)
- Universidad Politecnica de Madrid (UPM)
- Universidade Nova de Lisboa (UNL)
- University College Galway (UCG)
- Fraunhofer-Institut für Produktionsanlagen und Konstruktionstechnik (IPK)

Demonstrator B: Task Level Programming System

Demonstrator
Project Manager: Dr. Hörmann, UKA

Participants:
- FIAR SPA, Divisione Robotica Industriale
- Politecnico di Milano (PM)
- LADSEB-CNR
- Universität Karlsruhe (UKA)
- Gesellschaft für Prozeßsteuerung und Informationssysteme mbH (PSI)

Demonstrator C: High Level Interpreter

Demonstrator
Project Manager: Prof. Dillmann (UKA)

Participants:
- Universiteit van Amsterdam (UA)
- Universidade Nova de Lisboa (UNL)
- Universität Karlsruhe (UKA)

Demonstrator D: Planning and Interactive Programming System

Demonstrator
Project Manager: Mr Randet (Renault Automation)

Participants:
- Universität Karlsruhe (UKA)
- Renault Automation (RA)

Table 3.3:   Demonstrator project structure.

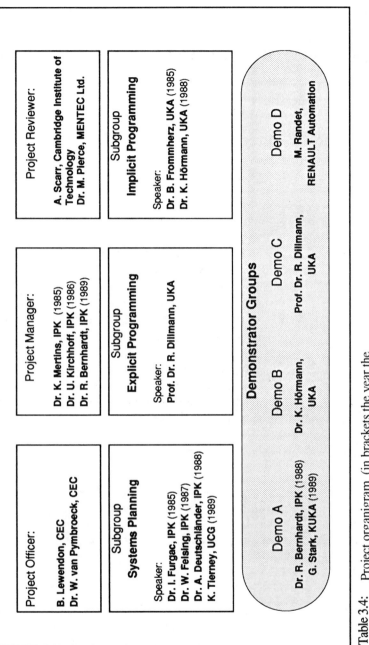

Project Officer:

B. Lewendon, CEC
Dr. W. van Pymbroeck, CEC

Project Manager:

Dr. K. Mertins, IPK (1985)
Dr. U. Kirchhoff, IPK (1986)
Dr. R. Bernhardt, IPK (1989)

Project Reviewer:

A. Scarr, Cambridge Institute of
Technology
Dr. M. Pierce, MENTEC Ltd.

**Subgroup
Systems Planning**

Speaker:
Dr. I. Furgac, IPK (1985)
Dr. W. Felsing, IPK (1987)
Dr. A. Deutschländer, IPK (1988)
K. Tierney, UCG (1989)

**Subgroup
Explicit Programming**

Speaker:
Prof. Dr. R. Dillmann, UKA

**Subgroup
Implicit Programming**

Speaker:
Dr. B. Frommherz, UKA (1985)
Dr. K. Hörmann, UKA (1988)

Demonstrator Groups

Demo A	Demo B	Demo C	Demo D
Dr. R. Bernhardt, IPK (1988) G. Stark, KUKA (1989)	Dr. K. Hörmann, UKA	Prof. Dr. R. Dillmann, UKA	M. Randet, RENAULT Automation

Table 3.4: Project organigram (in brackets the year the
responsibility was taken over).

Chapter 4

Structure of the book

R. Bernhardt
IPK Berlin, Germany

The structure of the book reflects to an extent the structure of the project itself. The contents have been subdivided into sections whereby Section II is dedicated to the planning system, Section III to the programming system, and Section IV to the information system. Section V describes various applications of the tools and methods developed within the project. Finally, a project survey is given in Section I, while Section VI contains a summary and an outlook on the future of robots in CIM.

In Figure 4.1 the section-oriented structure of the book is presented.

The sections are further broken down into chapters as shown in Figure 4.2. This figure should give the reader an overview of the contents of the book in one glance and also show the connections between sections and chapters.

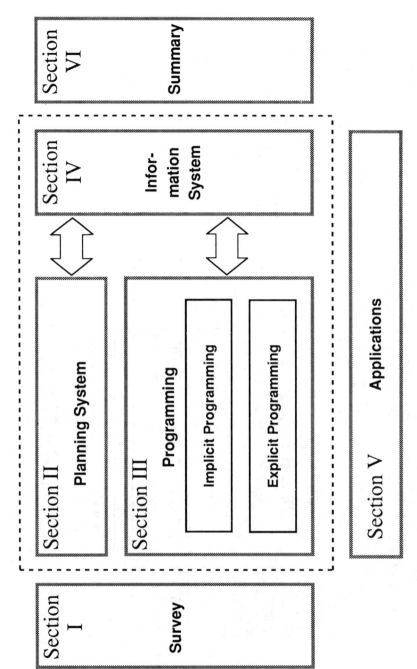

Fig. 4.1   Section oriented book structure.

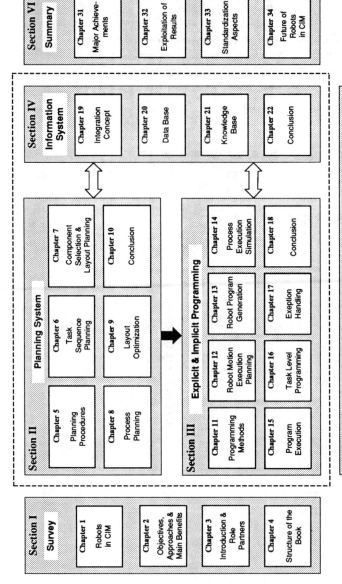

Fig. 4.2 Chapter-oriented book structure.

Section II

Systems planning

Chapter 5

Introduction to the systems planning process

V. Gleue
IPK Berlin, Germany

5.1 Objectives

The progress of automation technology has laid the foundation for the establishment of comprehensive production structures. Today's markets require increasingly the assembly of a variety of products in small lot sizes in optional sequence. To achieve economic returns on investment in spite of shorter product life cycles, flexible assembly systems are used. The extension of system performance requires a large number of parts, tools and joining materials to be placed at disposal. Therefore, future manufacturing systems will be characterized by a high degree of complexity.

The increased complexity of kinematics, material flow and communication increases the planning efforts required. Companies are forced to execute planning tasks more rapidly to put new products on the market as early as possible. Computer-aided planning tools offer the chance to plan more complex systems with a higher level of detail in shorter periods of time.

For the above reasons, a computer based system supporting the entire planning process has to be developed considering the following objectives:

- improvement of planning quality;
- reduction of planning costs;
- expansion of the spectrum of solutions to be able to make the best choices;
- increase in creative planning;
- fast processing and preparation of planning data;

- improvement of information flow;
- enhancement of the transparency of planning;
- establishment for robot integrated CIM-systems.

This is the subject of Section II of this book. As a first step to realizing a computer-aided planning system, a methodological procedure has been worked out taking into account existing models of industrial robot application planning systems. This procedure is presented in this chapter. The planning process is basically divided into seven phases (see Fig. 5.1):

- analysis;
- outline;
- documentation;
- design;
- evaluation;
- detailing;
- installation.

This procedure implies a successive and general approach. Due to the variety of assembly tasks in most cases it is necessary to select and connect the required planning functions to a suitable planning procedure. The planning steps are to be executed in an iterative manner, since in the course of planning assumptions made previously might have to be modified.

5.2 Phases of the planning procedure

5.2.1 Analysis

A detailed registration of the actual state of system and task is the basis and prerequisite for the planning process. This analysis is of high importance because its results determine the following planning steps. It contains the examination of spatial and temporal interaction of

- human beings;
- production means;
- work objects.

Relevant data and information for planning will generally be provided by information carriers such as manufacturing programs, drawings, part lists and working schemes and by studies on the scene.

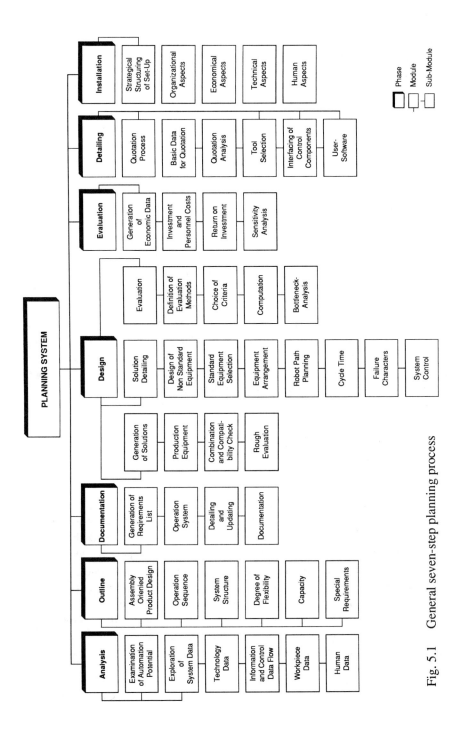

Fig. 5.1   General seven-step planning process

In order to estimate the necessary expenditure for automation by means of industrial robots, the central aim of the automation of handling processes must be to organize these processes as simply as possible.

5.2.2 Outline

In the outline planning phase, basic data and information have to be compiled to enable a precise formulation of the task. It is the purpose of this planning phase to rationalize existing processes and to point out all automation steps which have to be taken, such as the transition from manual to automated stage. In the process of planning new production plants, work procedure and production methods have to be determined at the first step, starting from geometrical and technical workpiece data. A restriction for automation is often a product design which is not assembly-oriented. According to the production task, appropriate production methods must be chosen from technical and economic considerations. In the next step the most suitable assembly sequence and manufacturing system architecture has to be determined. In conclusion, equipment and personnel capacities must be estimated.

On the basis of the planning work already done, the demand for manufacturing and testing equipment can be calculated. The determination of the required degree of automation and flexibility of a manufacturing system is particularly important. Based on the defined work sequence and system structure, the required area can be determined. In a subsequent step the main requirements of the control structure are defined. Finally, experimental work has to be done, especially in case of innovative technologies. It is one goal of the computer-aided planning to minimize efforts for experimental work.

5.2.3 Documentation

After defining and structuring the task, the process of detailed data preparation follows. The result of this descriptive information processing is the compilation of all demands and wishes in a list of requirements. Demands and wishes must be documented in figures as far as possible. Where this is not possible, they must be clearly expressed in verbal statements. In order to obtain a clear structure of the requirements profile, it is advisable to organize the list of requirements into fields of working, handling, placing at disposal, conveying, and others. This separation into special tasks is advantageous because it helps to avoid any loss of information in the later realization of the individual elements. This information documentation must be detailed and actualized as planning progresses.

5.2.4 Design

In the preceding planning phases the basic prerequisites for the system design have been worked out and formulated so that subsequently the selection and spatial arrangement of the single system components can be planned. The design phase is basically divided into the following tasks:

- development and selection of commercial manufacturing systems, feeding and orienting devices as well as conveying systems;
- determination of the basic kinematic concept of handling devices;
- elaboration of various equipment configurations;
- assessment of the concepts;
- determination of the number and kind of axes of the industrial robot;
- kinematic simulation;
- examination of the system behaviour.

It is advisable to select working systems, conveying media, feeding, orienting and other devices by means of describing the working process with standardized symbols. The qualitative description of the handling procedure can use the symbols for handling functions of the VDI-recommendation 2860.

In a subsequent step, a discursive method can be used to find principal solutions, e.g. the method of morphological matrices. The main task of the computer is to support the planner in his work combining the various functional means to a total solution. The positions of the necessary production means and feeding, orienting and testing installations are determined to a high degree by the basic concept of an industrial robot.

After determining the exact movements, handling times can be calculated on the basis of the performance data of specified industrial robots. After that the times of the machining systems can be determined and the cycle times of the total system can be co-ordinated. Since the application of automated manufacturing concepts generally requires high capital expenditure, it is necessary to test the operative quality of the planned system, so that investment risks are reduced.

5.2.5 Evaluation

In this step the most suitable system variant for an individual case of application is chosen by means of an evaluation procedure. Not only economic criteria expressed in money values have to be regarded. Social, human and technical aspects make the system evaluation a multi-dimensional decision problem.

In order to avoid mistakes and insecurities in evaluating system suggestions in the foregoing planning phase, it appears to be useful to evaluate several variants. For this reason, only absolutely inappropriate solutions should be removed from the group of possible solutions during the design phase. Considering several layout suggestions from an economic point of view increases the security of planning, for frequently variants which appear to be equally valuable regarding technical aspects may be characterized by large differences in capital expenditure.

5.2.6 Detailing

In this planning phase, documents are specified for system components which are not available on the market, and changes in existing production plants are planned. For instance, special devices or grippers have to be designed and built. If several systems have to cooperate to accomplish a task, existing manufacturing installations have to be adapted. For the interfacing of system components with the aim of synchronizing the operation sequence, data and signal communication must be realized.

5.2.7 Installation

In order to minimize unavoidable manufacturing interruptions for the introduction of new manufacturing installations, the temporal sequence of the system introduction must be exactly planned and surveyed. Production plants which are especially complex and require high capital expenditures are characterized by a high number of mutually dependent single processes and by many different processes which partially run parallel to each other. In such cases, the planning of a project can be improved by means of network analysis. With the help of this planning aid it is possible to shorten the execution time of a project because temporary critical proceedings can be recognized in time and appropriate countermeasures can be taken. Since during the introduction of a system generally a number of disturbances and delays occur, the temporal sequence must be corrected. Computer-aided planning means can minimize time and costs of producing and planning records. Furthermore, various measures against disturbances can be evaluated as to their effects.

5.3 Conclusion

The seven-stage iterative planning procedure just described forms the basis of a computer-aided planning system which has been developed under ESPRIT project 623. As the planning proceeds, the level of detail increases. Furthermore, the results of consecutive planning stages may be fed back to previous stages and cause iteration. Chapters 6 to 9 now present a number of methods and software tools which have been developed to address the key activities in the planning procedure, namely, task-sequence planning, component selection and layout planning, process planning and layout optimization. Chapter 10 summarizes and concludes Section II.

Chapter 6

Task sequence planning

K. Tierney, J. Browne
CIM Research Unit,
University College Galway, Ireland[*]

6.1 Introduction

The first step in planning is to generate basic data on how the assembly task can be performed and what equipment can be used to perform the task. The objective of this step is to identify a set of feasible solutions for the operation sequence plan and equipment selection. The results are stored in a relational database which can be accessed by the next planning step, where detailed planning will define the actual devices to be employed and the initial layout of the assembly workcell. Since the planning procedure [1] is iterative, the results of component selection and layout planning may be fed back to task sequence planning and cause replanning.

Task sequence planning for assembly tends to be time-consuming and is often still performed manually or with standalone systems. Lack of standardization can be a major problem which leads to a profileration of 'unique' assembly plans, many of which may be unnecessary. This task sequence planning problem for assembly may be characterized as follows [2]:

- No standard solution process is known apart from 'rules' and 'educated guesses' and there is rarely an optimal or unique solution. A solution technique which provides a reasonably good answer is as much as can be expected.

---

[*] The following people contributed to this research work in the CIM Research Unit during the life of ESPRIT Project No. 623: Richard Bowden, James Maguire, Maurice Walsh, Allen Moran

- In assembly, once certain precedence relations are satisfied, there are a number of equally feasible solutions.
- The problem is usually the responsibility of an experienced assembly planner much of whose knowledge is of a diffuse and heuristic nature.

The approach taken here was to develop a tool which would speed up the planning process and enforce standardization in assembly task sequence planning. The tool would also help to avoid the generation of different task sequences for similar products, by using a common knowledge base. Given the lack of scientific knowledge for assembly operations, the aim was to create a basic shell of an expert system which could gradually be built up with new knowledge. The concept was that if the assembly of a product could be understood in terms of standard operations [3] then the process/manufacturing engineer could develop feasible sequence of operations. To support this task sequence planning activity, the tool would also provide precedence analysis, as well as a pre-selection of feasible equipment to perform the assembly task. Finally the tool would be used in a decision support mode rather than as a device which automates the process planner's task, since it was considered that situations could arise where the recommendations of the system would not be acceptable to the user and would need to be over-ridden.

6.2 The task sequence planning tool for assembly

The tool described in the following sections consists of several modules: operation sequence planning, precedence analysis, robot pre-selection, gripper pre-selection and feeder pre-selection. Their interconnections are shown in Fig. 6.1. The tool consists of a rulesbase written in OPS5, an interactive spreadsheet written in FORTRAN-77 and and interface between the rulesbase and the spreadsheet written in BASIC. OPS5 is a language used in the field of artificial intelligence based on the production rule paradigm. It is a product of Digital Equipmnent Corporation. Each rule consists of two parts: (a) a series of conditions, and (b) a series of actions. The interpreter selects a rule to execute by attempting to match the first part of the rule (i.e. the condition) to the current data stored in the working memory. If the match is successful, then the corresponding series of actions is carried out. The software is implemented on a VAX 8700, under VAX/VMS operating system.

6.3 Operation sequence planning

In this module, the manufacturing or process engineer uses an interactive spread sheet as shown in Fig. 6.2. The spreadsheet displays the components of the product to be assembled (read from a Bill of Material file), a list of standard å operations (generated from case studies and from work done by Nevins *et al.* [3]), a user input facility and a message board. The user suggests the first å step and this is matched against the rulesbase which contains knowledge relating to *component design data, component orientation data, assembly operations databill of material data* and *gripper data*. Feedback is given via the message board to the user who may then modify the assembly step. This is repeated successively until a feasible sequence of operations for the assembly of the product has been built up and this sequence is then stored in the database.

Before an engineer can proceed in this fashion to generate an operation sequence plan for a product, data must first be input into the bill of material, computer-aided design, assembly operation, component orientation and gripper description data files. The rulesbase uses these data to fire rules and give feedback to the engineer. Initially a great deal of user input is required but once this data is present, any number of plans for a product can be generated. The data, rulesbase and results will now be described in more detail [4].

- **Bill of Material Data:** the bill of material data describes the structure of the product and the quantity of components per assembly.

- **Component Design Data:** Design data pertains to the relative shape, weight, size and surface contour of a component. Each component is assumed to have a maximum of six sides, and the data is used to recommend which sides of a component should be gripped.

- **Assembly Operations Data:** Assembly operations data consists of 'standard' operations and company specific operations. The 'standard' operations defined in this system are as follows:

  - insert screw;
  - peg in hole;
  - spot weld;
  - position;
  - push/clip;
  - flip over;
  - force fit;
  - remove;
  - rivet;
  - temporary support.

38

- **Orientation Data:** In the assembly process, a component may be involved in two distinct actions: the component is being assembled to something, OR something is being assembled to the component. The orientation data decides the most suitable orientation of a component depending on the situation. In the first case, an insertion face or mating face is identified, while in the second, particular features of the component are described which will specify a suitable orientation for a component e.g. hole, thread, chamfer. The feature is specified by the component side and corresponding axis and direction.

- **Gripper Data:** A library of grippers has been set up which can be updated continuously. The type of grippers include two-jaw scissors, two-jaw parallel, 3-finger, 4-finger, suction pads and electromagnets.

- **The Rulesbase:** The rules dealing with planning issues are grouped into sets to check and/or issue recommendations on the following features:

- fixture requirements;
- relative size of components being assembled in one assembly step;
- gripping face(s) of component with respect to surface contour;
- gripper type;
- number of tool changes between assembly steps;
- number of changes in direction of assembly between assembly steps;
- quantity per component/assembly constraints;
- orientation of components with respect to prominent features.

The particular rules which fire from within these sets depend on the geometry, orientation and surface contour of the components, and on the operation and assembly direction of each step, all of which are in some way specified by the user's input. A sample rule is as follows:

```
P Analyse_3.02A
    (goal ^ aim continue ^seek 2)
    (answer ^reply <reply1> ^opno <reply2>)
    part ^number <reply1> ^sc3 -flat- ^sc4 -flat-)
->
    (make counter ^sign 6)
    (make feed ^back Face 3 and Face 4 are the recommended faces to grip)
    (make chosen ^gripper two-jaw)
    (make face ^name1 3 ^name2 4)
    (modify 2 ^aim face)
```

(remove 2))

This rule recommends the faces on the product to grip, based on an analysis of the surface contour of the product i.e. if the surface contours of faces 3 and 4 of the first component to be assembled are both flat, then these faces are recommended for gripping and the recommended gripper is a two-jaw gripper.

- **Results of Operation Sequence Planning:** Fig. 6.3 is a sample of an operation sequence for the assembly of a telephone. For each assembly step, the output specifies the two components being assembled, the assembly operation, the axis and direction of assembly, the gripper to be used, the faces to be gripped, and whether a second robot arm is required.

6.4 Precedence analysis

The purpose of this analysis is to specify pre-condition operations and post-condition operations for each of the operations specified in the sequence planning module. This type of analysis assists the manufacturing or process planner in the task of assigning assembly operations to assembly stations. The analysis is performed via a rulesbase which checks the previous and subsequent operations to determine precedence relations. The rules cover the following situations which can occur during the assembly of a product:

- A component is placed on a fixture;
- Two components are assembled together;
- The assembly operation involves only one component or sub-assembly, for example a component being flipped over;
- The base part is removed from the fixture, implying that the assembly is finished [5].

6.5 Robot pre-selection

The purpose of this module is to aid the engineer in generating a list of feasible robots to perform the assembly task. This will effectively narrow the field of search for the user, but the final layout of the assembly cell will be the most important consideration. The technical parameters used in generating the list of feasible robots include:
- repeatability of the robot arm;
- robot work envelope;
- robot programming method;

- robot load capacity;
- robot drive method;
- number of degrees of freedom.

The first five parameters are specified by the user and the number of degrees of freedom is calculated from the assembly plan. A series of rules, which contain information on a library of robots, is used to identify suitable robot types. This list is then stored in the relational database for use at the next planning step.

6.6 Gripper pre-selection

The purpose of the gripper selection module is to give a detailed description of the gripper specified in the results of the operation sequence planning module. Based on geometric features of the components in the assembly and on user-specified parameters (e.g. reliability), the rule base develops an outline specification for the gripper. This specification is stored in a relational database.

6.7 Feeder pre-selection

The purpose of the feeder selection module is to select an appropriate feeder for a component, based on the component's design/geometric characteristics. The selection is achieved by numerically encoding components. The coding scheme is expressed as a series of rules. The first digit of the code is determined by the general shape of the component and the ratio of its dimensions, which are supplied by the user. The user then answers a series of detailed yes/no questions on the components design features, by which means values are assigned to the second and third digit of the code. The code is then matched against a library of feeders in the database and a set of suitable feeders are selected. Several of the feeding devices are capable of feeding a wide range of components, e.g. vibratory bowl feeders.

6.8 Conclusion

The tool described in this chapter generates basic data on how a product can be assembled and the likely range of equipment which can be used to perform the task. This tool is a prototype development which is continually being revised by encoding new knowledge of robot-based assembly and by developing plans for different products, in order to enrich the rulesbase. OPS5 supports this process since new information can be incorporated into the tool in the form of production rules, with little change, if any, to existing rules.

The results of this task sequence planning step are stored in a relational database to be accessed in the next planning step where individual devices are chosen and the assembly workcell layout is planned. Since the choice of a particular robot or piece of equipment may entail specific assembly requirements, the operation sequence plan may need to be refined on the basis of this new, more detailed information. Thus the planning procedure is iterative.

Component selection and layout planning will be described in the next chapter.

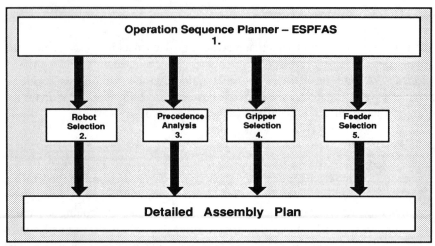

Fig. 6.1   Assembly sequence planning system.

Part No.	Operation Sequence	
Level 2 16 PCB ASSY 1 17 K/BOARD 1 18 SCREEN 1 19 BASE 1 20 GEN.ACCTIC 1 21 ADHESIVE 1 Level 1 5A B.T.HARMNY 1 ALTR SWTH 1 2 RBBR FEET 4 3 SMALL WT 1 4 LRGW WT 1 5 N/WRK DL 1	Please input the first component being assembled (e.g. 1-,2-,3- etc.) ->>	Message Board Insert S to scroll

ESPRIT Project No. 623	Assembly Operations			
	A INSERT_SCREW B PEG_IN_HOLE	C SPOT_HELD D POSITION	E PUSH/CLIP F FLIP_OVER	G FORCE_FIT H REMOVE

Fig. 6.2   The spreadsheet.

ASSEMBLY STEP		ASSEMBLY OPERATION	AXIS & DIRECTION OF ASSY		GRIPPER	FACE	2ND ROBOT ARM?
PCB_ASSY	TO FIXTURE	VIA POSITION	Y	N	SENSOR	3 4	N
KEYBOARD	TO FIXTURE	VIA POSITION	Y	N	TWO_JAW	5 6	N
KEYBOARD	TO PCB_ASSY	VIA SPECIAL	X	N	****	****	***
SCREEN	TO PCB_ASSY	VIA POSITION	Y	N	SUCTION_PADS	1 0	N
KEYBOARD	TO ITSELF	VIA FLIP_OVER	Y	P	TWO_JAW	5 6	N
KEYBOARD	TO DUMMYD1	VIA PUSH_CLIP	Y	N	TWO_JAW	5 6	N
N/WORK_DL	TO FIXTURE	VIA REMOVE	Y	P	TWO_JAW	5 6	N
BASE	TO FIXTURE	VIA POSITION	Y	N	TWO_JAW	3 4	N
ACCOUSTIC	TO BASE	VIA PEG_IN_HOLE	Y	N	SUCTION_PADS	1 0	N
DUMMYD2	TO ITSELF	VIA SPECIAL	Y	N	***	****	***
BASE_ASSY	TO FIXTURE	VIA REMOVE	Y	P	TWO_JAW	3 4	N
BASE_ASSY	TO FIXTURE	VIA POSITION	Y	N	TWO_JAW	3 4	N
SMALL_WGHT	TO BASE_ASSY	VIA PUSH_CLIP	Y	N	TWO_JAW	1 2	N
LARGE_WGHT	TO BASE_ASSY	VIA PUSH_CLIP	Y	N	TWO_JAW	1 2	N
SPRING	TO BASE_ASSY	VIA PEG_IN_HOLE	Y	N	SUCTION_PADS	1 0	N
PLUNGER	TO BASE_ASSY	VIA PUSH_CLIP	Y	N	SENSOR	1 0	N
ALERTER	TO BASE_ASSY	VIA PEG_IN_HOLE	X	N	ORIENTATION	1 0	N
LABEL	TO BASE_ASSY	VIA SPECIAL	Y	N	***	****	***
N/WORK_DL	TO DUMMYD3	VIA PUSH_CLIP	Y	N	TWO-JAW	5 6	N
N/WORK_DL	TO DUMMYD3	VIA SPECIAL	Y	N	***	****	***
HOUSING	TO DUMMYD4	VIA PEG_IN_HOLE	Y	N	TWO_JAW	3 4	N
DUMMYD5	TO ITSLEF	VIA FLIP_OVER	Y	P	TWO_JAW	3 4	N
FOOT	TO DUMMYD5	VIA PUSH_CLIP	Y	N	SUCTION_PADS	1 0	N
FOOT	TO DUMMYD5	VIA PUSH_CLIP	Y	N	SUCTION_PADS	1 0	N
FOOT	TO DUMMYD5	VIA PUSH_CLIP	Y	N	SUCTION_PADS	1 0	N
SCREW	TO DUMMYD5	VIA INSERT_SCREW	Y	N	TWO_JAW	1 0	N
SCREW	TO DUMMYD5	VIA INSERT_SCREW	Y	N	TWO_JAW	1 0	N
SCREW	TO DUMMYD5	VIA INSERT_SCREW	Y	N	TWO_JAW	1 0	N
CORD	TO DUMMYD5	VIA PUSH_CLIP	Y	N	SUCTION_PADS	1 0	N
CABLE	TO DUMMYD5	VIA PUSH_CLIP	Y	N	SUCTION_PADS	1 0	N
CORD	TO DUMMYD5	VIA SPECIAL	Z	P	***	****	***
CABLE	TO DUMMYD5	VIA SPECIAL	Z	P	***	****	***
DUMMYD6	TO ITSELF	VIA FLIP_OVER	Y	P	TWO_JAW	3 4	N
HADSET	TO DUMMYD6	VIA PUSH_CLIP	Y	N	----	---	N
HADSET	TO DUMMYD6	VIA POSITION	Y	N	----	---	N
NUMBER_CD	TO DUMMYD6	VIA POSITION	Y	N	SUCTION_PADS	1 0	N
WINDOW	TO DUMMYD7	VIA PUSH_CLIP	Y	N	SUCTION_PADS	1 0	N
B.T.HARMNY	TO FIXTURE	VIA REMOVE	Y	P	----	---	N

Fig. 6.3 An example of the output file.

References

[1]   Spur G., Furgac I., Deutschlander A., Browne J., and O'Gorman, P., Robot Planning Systems, J. of Robotics and Computer Integrated Manufacturing, 2 (1985) 115.

[2]   Kempf, K. G., Manufacturing and Artificial Intelligence, Robotics I (Elsevier, 1985), p. 13-25.

[3]   Nevins, J., and Whitney, D., Computer Controlled Assembly, Sci. Amer., 238 (1978) 62.

[4]   Tierney, K., Bowden, R., and Browne, J., ESPFAS - A Prototype Expert System for Technological Planning in Robot-Based Flexible Systems, Annals of Operation Research, 15 (1988) 111-130.

[5]   Bowden, R. and Browne, J., ROBEX – An artificial intelligence based process planning system for robotic assembly, IXth Int. Conf. on Production research, Cincinnati,, (1987).

Chapter 7

Component selection and layout planning

V. Gleue, S. Krüger
IPK Berlin, Germany

S. Wadhwa, K. Tierney, J. Browne
CIM Research Unit
University College Galway, Ireland[*]

7.1 Component selection

7.1.1 Introduction

The equipment of automated assembly systems can generally be classified into two categories:

- standard task-independent components and
- task-specific components.

Examples for standard components are industrial robots, conveyor systems and sensors. Task-specific equipment like endeffectors, fixtures and feeding components have to be added as required by the respective assembly task. To increase the degree of reuseability of an assembly system in case of product changes, the number of task-specific components should be minimized.

---

[*] The following people contributed to this research work in the CIM Research Unit during the life of ESPRIT Project No. 623: Thomas O'Donnell, Feargal Timon, Padraig Bradley

The selection of the most suitable standard components on the market as well as the design of grippers, feeders and fixtures is in most cases a time consuming task which extends planning periods and raises project costs. On the other hand, great care should be bestowed on this task because the equipment influences mainly the costs per product unit and causes high investment expense. For the design of task-specific components, paths and product geometry have to be considered. Additionally, associated components must be taken into account in order to avoid collisions and achieve short task times. To fulfill this requirement, prototypes and, subsequently, component modifications are necessary.

The developed planning tool intends to support the planner in selecting and designing components by

- supplying the planner with a standard components library and
- visualizing gripping and joining processes.

The basis of the tool is a set of three-dimensional CAD models of all components and an advanced graphic workstation which allows the models to be manipulated on-line.

7.1.2 Basic instruments of the developed tool

The basic instruments of the developed planning tool are the three-dimensional CAD systems COMPAC and ROMULUS. To graphically manipulate the models a workstation PS 390 from Evans & Sutherland is used.

All product parts and cell equipment in the planning system are modelled with the 3D-CAD System COMPAC (COMPUTER ORIENTED PART CODING). It was developed at the IPK Berlin. In the following a short description of COMPAC is given.

Since most workpieces in the area of mechanical engineering are enveloped with planar, cylindrical or conical faces they can be considered as composed of blocks, cylinders and truncated cones. To model a workpiece these primitives are used. They are dimensioned, positioned and added to or subtracted from each other. To facilitate modelling closed contours can be defined and, for example, can be rotated around an axis. Parts with irregular faces, e.g. car body, can be created by spline-curves. Figure 7.1 shows the construction of components using the COMPAC primitives.

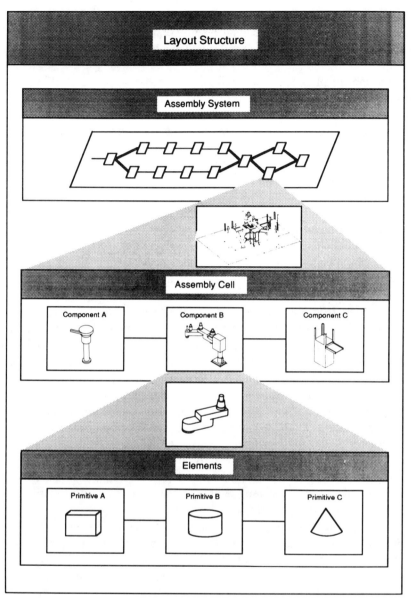

Fig. 7.1   Primitives for modelling in COMPAC.

The system generates a three-dimensional computer-internal model of the workpiece. This model serves as a basis for the following planning steps layout planning, process planning and simulation since its geometrical data can be analyzed and processed by the computer. All workpieces and devices involved in a given assembly task are modelled and stored in a relational data base (see Chapter 20).

For graphical planning and simulation tasks the workstation PS 390 from Evans & Sutherland is used. It allows the user to create, display, modify and manipulate complex 2D and 3D wireframe models interactively. The user can move, scale and rotate the images by any amount in any direction. Interactive devices are dials, key and mouse. They send values through user-defined function networks to interaction points in the model's structure. So the system can computer-internally represent kinematic configurations internally and allows their movements to be simulated.

7.1.3 Design of task-specific components

There are close relationships between the task-specific components of an assembly systems. For example, outlining a gripper requires knowledge not only about the workpiece to be grasped but also about the geometrical conditions of joining and taking the part from the presentation. On the other hand, the designer of part presentations and fixtures has to regard the gripping method and the size of the gripper. The developed tool aims at shortening the planning and implementation period of task-specific components by visualizing their interactions on screen and transferring the test and verification into the conceptual phase.

The system supports the planner in determining the optimal gripping faces of a part. Different orientations between gripper and workpiece can be analyzed quickly on screen. The CAD models of part and gripper are loaded to the workstation PS 390. The gripper is classified as 'moveable' and the part as 'fixed'. The dials are connected to the moveable model so that the user is enabled to choose the most suitable gripping faces of the part and the best orientation of the gripper to pick. As the last step the chosen frame of the part in relation to the gripper is stored in the database.

To analyze picking and joining conditions the same tool function is used. Figure 7.2 shows different strategies for gripping a dc-actuator from its presentation. The gripper, workpiece and parts presentation are displayed. While moving the gripper and part using the dials the graphic features of the PS 390, such as scaling and rotating the models, enable the user to identify geometrical restrictions. Gripper or part´s presentation design modifications can be inferred. Additionally, the system allows the planner to determine picking and joining paths. They are stored in the data base and can be used in the following process planning phase.

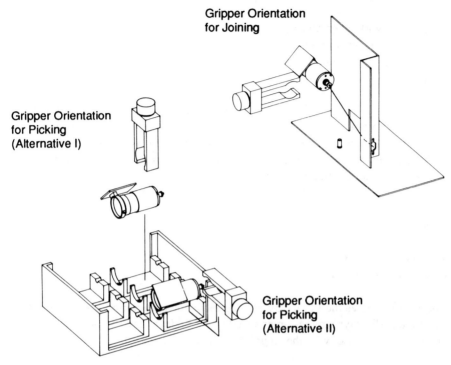

**Gripper Orientation
for Joining**

**Gripper Orientation
for Picking
(Alternative I)**

**Gripper Orientation
for Picking
(Alternative II)**

Fig. 7.2 Strategies for picking a DC-actuator from its presentation.

7.1.4 Selection of standard components

The selection of cell equipment is a two-step process. Based on the assembly task to be executed a layout-independent pre-selection has to be carried out. It has to regard criteria like

- minimum load;
- kind of actuation;
- control;
- programming;
- ability of informational linkage.

This is done in the previous planning phase (see Chapter 6). The pre-selected components are stored in the data base.

In the second step criteria depending on the topological arrangement of the components have to be regarded. These are e.g.

- required working envelope;
- number of gripper axes;
- rated load (part load plus tool load);
- speed.

To support the process of selecting equipment a library of standard components has been implemented. It represents the first step towards a tool which is to consult the planner by means of rules in a later development phase.

In the current version the library is a structured data base. The standard components are classified in the categories

- robots;
- conveyors;
- feeders;
- storage devices/tables;
- sensors/testing units.

The planner views the library by picking the name of a component (Figure 7.3). The system displays a three-dimensional CAD model of the component on screen and provides the user with the technical data of the component, e.g.

- working envelope;
- number of axes;
- minimal load;
- speed;
- kind of control.

The user is enabled to select components for the layout by picking the displayed name of the component or the graphic model. The number of times the planner picks, the component will appear in the layout. It is also possible to cancel an already selected component.

In the next step the peripheral devices have to be chosen. Main tasks are handling, especially ordering, and storage functions. Similar to robot grippers the feeding and storing devices are often workpiece specific. There are only a few standard peripheral components on the market, since they cannot achieve the flexibility of a robot.

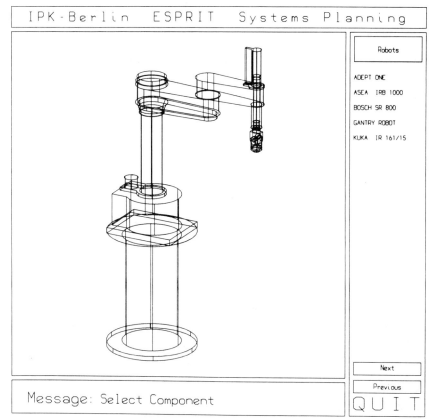

Robots

ADEPT ONE

ASEA IRB 1000

BOSCH SR 800

GANTRY ROBOT

KUKA IR 161/15

Next

Previous

QUIT

Message: Select Component

Fig. 7.3   User menu for component's selection.

7.2 Layout planning

In this planning phase the arrangement of the cell components has to be elaborated. There are two steps to this phase. In the first step, the main objective is to minimize the paths of the robot as well as the kinematic requirements. This depends mainly on the working envelope of the chosen robot. The second step evaluates the layout from a material flow perspective in order to determine the control strategy required to fulfill a given material flow requirement, e.g. cycle times and their variability with regard to the flow of assemblies through the cell.

Starting from the manipulator's kinematic system the planner finds the principal structure of the component's arrangement. For instance, for a robot with a cylindrical or spherical work envelope an ordering of the equipment in a circle would best fit with the layout planning objectives. Based on this first approach the

51

planner is able to improve and optimize the arrangement. It should fulfil the following requirements:

- loading and un-loading the components without using the robot´s hand-axes;
- minimized number of points to approach between two components;
- minimized distances between components sequentially to approach.

Restrictions result from area available, material flow in the factory etc. Additionally, component specific restriction like

- load and un-load conditions;
- required area for servicing and repair;
- existing machine foundations.

often reduce the possibility to design an optimal cell layout.

The developed system places the selected equipment (see Chapter 7.1) on a virtual shop floor represented by a displayed grid. To enhance the transparency of the geometric conditions the grid has a one meter scale. The planner can now stepwise arrange the equipment on the shop floor. The components are connected to the dials and can be moved and rotated on-line. At any time the planner can switch back to the data base and add or delete components.

In spite of the possibility to change freely the viewpoint or to scale the displayed cell model it may be difficult for the planner to have a clear look at the layout due to the use of wire frame models, especially in the case of complex assembly systems. To enhance the transparency of the layout it is possible to remove components not of interest at the moment from the screen. They can be re-displayed at any time.

To reduce routine operations in the planner's work it is intended to support the planner in arranging the components. The system calculates a configuration and proposes a layout on screen to the planner which he subsequently can interactively modify and optimize. This program starts from the robot's envelope. Figure 7.4 shows the method of computer-aided arrangement of the components in the case of a SCARA (Selected Compliance Assembly Robot Arm) kinematic. The workspace is segmented according to the number and the shape of the peripheral components. The segments marks the suggested location for picking parts from a feeding device and executing mating processes. In order to reduce possible collisions between robot and peripheral devices the components should be arranged so that only a defined work area is located within the work envelope of the robot. The result is a first rough cell layout. The second step in layout planning is to evaluate the design with respect to material flow parameters. This is done using material flow simulation. Unlike graphical three-dimensional simulators for the design of robot cells which focus on the detailed evaluation of a particular robot for a certain application, material flow simulation evaluates the layout design with respect to assigned operation sequences and designed control strategies. This results in the estimation of realistic cycle times for various design alternatives [1].

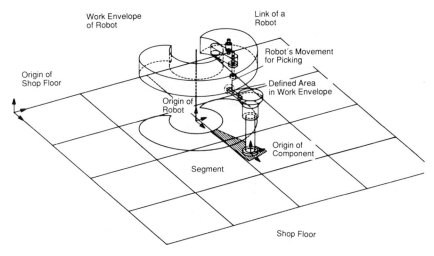

Fig. 7.4   Strategy for computer aided layout planning.

Two major constraints on the use of material flow simulation have been the availability of simulation model-building expertise and the long lead times required to develop detailed simulation models. To overcome these obstacles, a generalised data-driven material flow simulator has been developed to model a single robot cell.

This simulator, called FACET (Flexible Assembly Cell Evaluation Tool), is coded in the simulation language SLAMII and its user interface is supported by FORTRAN and SMG routines on VAX 8700, under VAX/VMS operating system.

7.2.1 FACET input

FACET allows the user to define the static configurations (layout, equipment, sensors), the assembly requirements (precedence relationships), gripper requirements, special processes (inspection, soldering, etc.), the interaction control (control strategies), the reliability parameters (feeder jams, robot breakdowns, etc.), and the operation control for various processes (component

53

feeding, robot flow, workpiece flow, fixture flow). This is accomplished by guiding the user through an interface built with a number of appropriate menus. An analysis of the operation of a number of assembly cells indicates that with these definitions, there exists a one-to-one correspondence in the logical relationships for various processes in the cell. Thus, FACET simulation code is a mapping of these logical relationships from the definitions of the cell design and operation [2].

Figure 7.5 shows the sample data input screens from FACET. The top screen presents a summary of the current setup of the cell. The bottom screen illustrates how the designer is aided in entering the operation sequence with respect to the order of assembly of the components. Based on the gripper requirements for each component, FACET prompts with an effector change indication between the components. Such a screen is presented for each product type to be assembled in the cell.

FACET - THE CURRENT SETUP

Fee der	Part Name	Part No.	Presence Detector	Pre-orient	Vision	Pallet	Start Buffer	Max Buffer
1	Cmp1	CP/1	Yes	No	Yes	No	0	5
2	Cmp2	CP/2	Yes	Yes	No	No	0	5
3	Cmp3	CP/3	Yes	No	Yes	No	1	5
4	Cmp4	CP/4	Yes	Yes	No	No	0	5
5	Cmp5	CP/5	Yes	Yes	No	No	0	5

Grippers/Tools

Suction Gripper	Magnetic Gripper	Vacuum Gripper	3-finger Gripper	Solder Tool	Inspect Tool
External	External	External	External	No	No

COMPONENTS		Assembly Precedence	SEQUENCE
Part name	Required		Cmp1
			Gripper change
Cmp5	1	Product : PROD-1	Cmp2
Cmp4	1	Part no.: MP/1	
		Use select key after movimg arrow to the component row.	

Fig. 7.5   A sample input data screen from FACET.

7.2.2 FACET output

The material flow performance statistics generated by FACET consist of the cycle time characteristics, utilisation level of equipment (grippers, tools, feeders, robot etc.) and the interaction delays due to the lack of synchronisation on the interacting processes. The cycle time evaluation is the primary objective of FACET. If the cell is a stand-alone unit, the cycle times are a measure for prioritising the design alternatives. The utilisation and delay statistics are normally used for the design modifications. At the system level (i.e. when the designed cell is part of a system) the cycle time characteristics may be important depending on the type of the system configuration.

Figure 7.6, a typical result from FACET, illustrates that a mean cycle time of 133 time units can be expected for a product. It also shows that 61% of the products will have a cycle time below 125 time units and 87% will have cycle times below 176 time units. The variation is caused by various breakdowns and the effect of various process interactions in the cell.

Total No of Obs		90		
Average Value:		492		
Lower	Upper	Obs	Freq	Cum
483	493	76	0.84	0.84
494	502	1	0.01	0.85
503	512	5	0.05	0.91
513	522	0	0.0	0.91
523	531	0	0.0	0.91
532	541	1	0.01	0.92
542	550	1	0.01	0.93
551	560	0	0.0	0.93
561	570	0	0.0	0.93
571	579	6	0.06	1.0

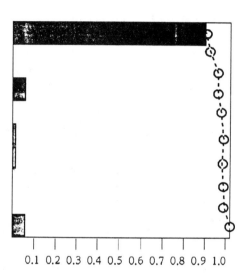

0.1 0.2 0.3 0.4 0.5 0.6 0.7 0.8 0.9 1.0

Fig. 7.6   Cycle time characteristic for a robot assembly cell generated by FACET.

7.2.3 FACET modelling framework

The basic features of the operation of any single robot cell are the robot flow process, i.e. the controlled flow of the robot to travel to and from the various task areas to accomplish such tasks as component pickup, assembly and tool changes, and its interaction with other flow processes such as component flow, information flow, tool flow, base assembly flow and fixture flow. At each interaction a robot task is exercised. For cell design purposes, the robot motions are actually divided into two components:

1. *A path motion*: the gross motion period of the robot. Characterized by high velocity and low accuracy motion, it is used for travelling between the various task areas. The times taken for path motions are considered to be deterministic.
2. *A task motion*: an interface motion and a fine motion. The interface motion brings the robot to the task surface. The fine motion accomplishes the assembly task. These motions are relatively slow and accurate. The times associated with task motions in assembly are usually stochastic, especially when compliant devices are in use [4].

From a modelling perspective, the cell is viewed as a framework of task areas. The robot visits these task areas according to a planned operation sequence. This framework reflects the layout of the cell in terms of the relative locations of the equipment (feeders, tool changer, fixture) associated with each task area. The layout is quantified by the times taken to travel between the task areas. Each task area is quantified by the task time depending on the task to be accomplished there. Each task area is associated with an interacting process, for example, the component C1 pickup task area is associated with the feeder process corresponding to component C1. A control or decision point is defined at each task area [3]. If the task has been accomplished, this decision point determines the next destination of the robot according to the operation sequence. If the robot has just arrived to accomplish a task, the decision point determines whether the robot needs to be delayed, oriented, or allowed to proceed with the task. The nature of decision making at each decision point is inherited from the design and control of the cell.

From the cell simulation perspective, the robot flow in the cell is modelled as the time relationships of the robot travel and the robot tasks in the above-described framework. The logical relationships of the robot flow process are inherited from the decision points for each task.

7.2.4 Material flow simulation in other domains

The principles on which FACET are based have been used to develop two other simulators, ROBSIM (ROBot SIMulator) which models the operation of a multi-robot assembly cell and FASIM (Flexible Assembly SIMulator), which models the operation of a flexible assembly *system*, consisting of a number of assembly cells, a transport system, auxilary cell (e.g. inspection, repair, carousel) and system control strategies [2]. Furthermore, an expert system has been built for FASIM in order to analyze simulation output and make recommendations for system modification [5], [2].

7.3 Conclusion

This chapter describes tools which have been developed to provide decision support in the selection of component devices to be used in a robot-based assembly cell and in the layout planning of the cell. This planning step generates more detailed data than the previous planning step, and therefore the results are fed back to task sequence planning to refine the operation sequence plan and equipment selection which were initially outlined.

In the next planning step, the design of the robotic assembly cell is brought to a further level of detail whereby the components arrangement has to be analyzed and optimized with respect to criteria such as:

- minimal path length to move by the robot;
- no danger of collision;
- enough room for material supply;
- enough room for service and reparation purposes.

References

[1] Grant F.H., and Wilson, J.R.: 'Material Flow Evaluation in Robotic Systems using Simulation', Mater. Flow 3, 83-87, 1986.

[2] Browne J. and Wadhwa, S.: Simulation: Layout of a Robot Cell, in Robot Technology and Applications, ed. by U. Rembold, Marcel Dekker Inc., pg. 207-276, 1990.

[3] Wadhwa S., and Browne, J.: 'Modelling FMS with Decision Petri Nets', Inter. J. of Flexible Manufacturing Systems, Kluwer Academic Publisher, Vol. 1, 225-280, 1989

[4] Wilhelm W.E., 'An Approach for Modelling Transient Material Flows in Robotised Manufacturing Cells', Mater. Flow 3, 55-68, 1986.

[5] Wadhwa S., and Browne, J.: 'A Goal Directed Data-Driven Simulator for FAS Design', paper delivered at European Simulation Multiconference, July 8-10, Vienna, 1987.

Chapter 8

Process planning

V. Gleue, S. Krüger
IPK Berlin, Germany

8.1 Introduction

At the interface between the planning and programming tools, a computer aided assembly process planning tool has been developed. It is to comprehend and transform the elaborated planning data into a suitable form for the off-line programming process. The module gives as results a detailed operation sequence, the definition of robot paths as well as the determination of technological and logical conditions.

The basic for the planning of the assembly process is the decomposition of the entire assembly process into

- task;
- operation;
- element (see Fig. 8.1).

The task defines the work extent of the assembly process. To meet the task, a certain number of operations are executed in the system. An operation is a sequence of work elements and results in progress towards assembling the product. Elements are defined as constituents of an operation which can neither be decomposed in terms of its description nor its temporal measurement.

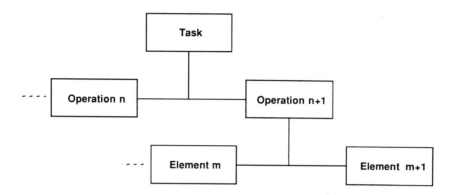

Fig. 8.1   Decomposition of an assembly task.

The process planning is based on the results of assembly sequence and layout planning. As the first step, a textual description of the assembly task is prepared which encompasses all the operations to be executed. Then the single operations are decomposed into their elements. Next they are attached to necessary data such as coordinates of robot paths or transmitter and receiver of signals. Finally, a task description in standardized form is generated which represents a first step towards the development of a robot program. Information transfer between the program system, the data base, a CAD system and a graphic system is a precondition for computer aided realization. Fig. 8.2 schematically represents the information input from previous planning steps as well as the procedure of planning the assembly process.

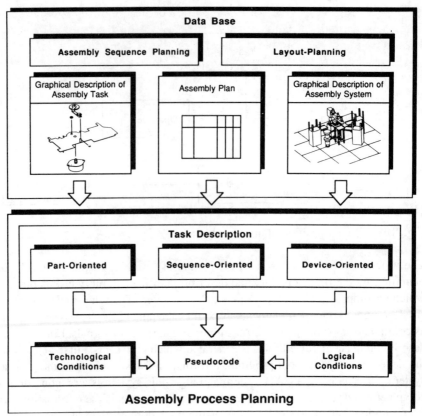

Fig. 8.2   Reference model of process planning.

8.2 Task description

The planning system has to provide a detailed assembly task description for the off-line programming phase. This demands planning information on

- parts;
- operation sequence;
- equipment;

which has mainly been elaborated in the preceding assembly sequence and layout planning phases. Accessing the relational data base, the system provides a table to the planner which contains information about the logical sequence of operations,

60

the joining parts, their relations to their mating parts and equipment as well as the joining technology (Fig. 8.3). Either a draft of the product exploded in its component parts or the related layout is presented on the graphic display to enhance the transparency of the task to be executed.

The second step of the process planning encompasses two tasks: on the one hand the elaboration of detailed planning data like coordinates of the robot paths and technological and logical conditions - e.g. the specification of speed or the transmission of signals - on the other hand the description of the assembly process in standardized form. Therefore, a pseudocode with commands and identifiers is used.

At first, each operation of the assembly task considered is decomposed into its elements. From the data base the user is provided with a mask in tabulated form which contains attributes concerning the parts, robot paths, communication between the cell components etc. (see Fig. 8.4). The planner completes the provided table by

- data acquisition from the graphic workstation;
- dialogue with the data base;
- user input.

Task : Drive Unit

Opera-tion No.	Part	Mating Part	Device				Joining Technology
			Part Presentation	Gripper	Clamp	Tool	
1	Gear	Side Plate	Vibratory Bowl Feeder	-	-	-	Assembling
2	.	.	.	.	.	.	.
3	.	.	.	.	.	.	.
.	.	.	.	.	.	.	.
.	.	.	.	.	.	.	.

Fig. 8.3   Table of operations.

8.3 Determination of the process

Using the graphic workstation PS 390 from Evans & Sutherland, the system enables the planner to simulate successively the entire assembly process on the screen. By visualising stepwise the sequence of single elements, the interaction between the cell components can be recognized and the planner is able to determine at which time certain components have to exchange signals. To identify the robot paths in the layout the system provides

a)   the possibility to move the graphic model of the robot manually with the dials of the graphic workstation and

b)   CAD based functions to identify frames which the robot approaches automatically.

To determine path coordinates which do not require high accuracy, the planner is able to move the robot joints by the dials of the graphic system and store the actual effector position in the data base. This procedure is similar to 'teach-in' programming methods.

Task : Drive Unit

Operation No. : 12

Element No.	Subject	Object	Action	Path	Speed	Signal	Receiver
1	Robot A	-	move	P_1	100	-	-
2	Robot A	-	move	P_2	30	-	-
3	Robot A	Gripper A	fix	-	-	-	-
4	Robot A	-	-	P_3	30	-	-

Fig. 8.4   Table for task description with standardized elements.

Using the CAD-based functions, the planner is able to identify exactly special coordinates to which the robot has to move. This identification should be carried out either in the process of designing the part-specific equipment or within the assembly sequence planning. By using the CAD system, the planner defines exactly the geometrical connection of related graphic models, for example between gripper and mating part. An example is shown in Fig. 8.5. As the relations between different graphic models change during the assembly process, all possible combinations have to be described and stored in the data base. This enables the program, for example, to transform the frames of the graphic models into the robot dependent frame.

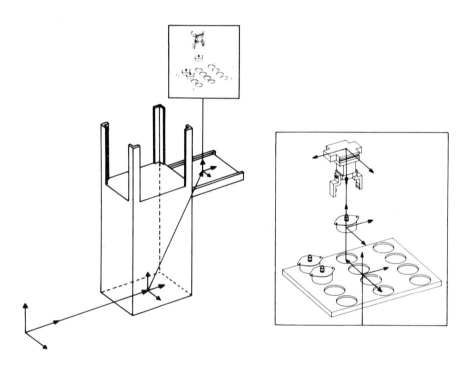

Fig. 8.5   Relations between components for planning of pick-up movements.

Based on an internal kinematic representation of the robot, the system is able to calculate the required configuration of the robot joints to approach the specified position and orientation. Significant robot positions are described as a frame, transmitted and stored in the data base. By holding the status of all processes throughout according to the assembly sequence, the whole process can be represented depending on time. Fig. 8.6 shows the system's user interface. Now that the detailed process plan has been generated, the next and final step in the planning procedure is to optimize the layout.

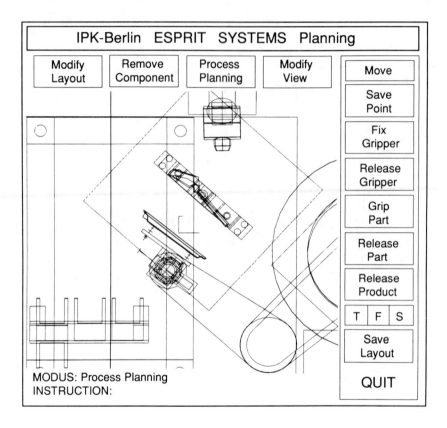

Fig. 8.6   User interface for assembly process planning.

Chapter 9

Layout optimization

G. Stark
KUKA Schweissanlagen & Roboter GmbH, Germany

9.1 Introduction

Optimization is an important requirement for the planning task. It is the final step within the system planning procedure. The global objective is to finally improve the characteristics of the system. Thereby, two aspects will be considered. The performance of the system has to be improved with respect to the task to be executed. General features of the system are to be optimized, too. This applies, for instance, to economical aspects like overall costs or customer preference for certain components to be used. Also general technical aspects such as covered shopfloor area have to be considered.

'Layout optimization' is an integral part of the overall planning system. Fig. 9.1 depicts a data flow diagram that shows its integration into the overall system. In particular there is an exchange of data with three adjacent planning modules:

- component selection and layout planning;
- robot execution planning;
- process execution simulation.

The 'component selection and layout planning' module supplies a draft of the layout of the manufacturing system that is to be planned. This covers a proposal of the components that have to be selected and a first definition of their location within the workcell. This information is used as an initial state for the optimization process. Additionally, there has to be information defining the objectives that have to be achieved by applying optimization.

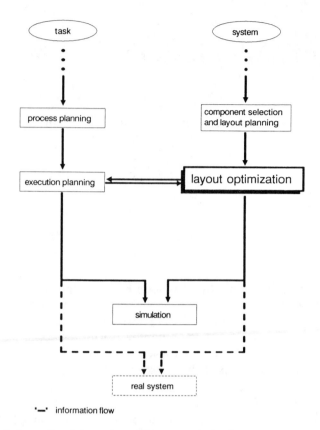

Fig. 9.1   Information integration of layout optimization.

The result of the optimization process is made up by the final specification of the components and its spatial and functional integration into a workcell that meets the requirements. This result will be supplied to the 'process execution simulation' module. 'Layout optimization' is closely related to 'robot execution planning'. The task that has to be executed is an essential input for the system optimization process. Therefore, an iterative exchange of information with the 'robot execution planning' module has to take place.

In particular, optimization support is a requirement for rather complex manufacturing systems. The features listed below are typical.

- There are many design variables to be decided (an approximate number is between 30 and 50).

- The real systems are subject to continuous change during the planning phase. The models representing them have to keep step with this process.
- There are many dependencies between the design variables. These are subject to rapid and frequent change, too.
- The dependencies between the design variables do not always obey mathematical laws. Often they are rather arbitrary because of the many internal and external influences which affect the system planning phase.

Following an industry-oriented approach, the main strategy has been to develop aids that can be put into practice before long. The experience gained in that way can be used to carry on further development efficiently.

A precondition of the application of computer aided techniques always is to represent properly the information that has to be processed. Recent applications in industry show that the representation of the constraints that govern the design variables is a very big and important problem. It is a precondition for being able to optimize the layout of complex manufacturing systems.

Practical applications show that the amount of data that has to be processed is huge. Therefore, a transparent and flexible representation of the models is an important demand. Particularly, this means that the parts of the models that are subject to frequent alterations have to be separated from invariant parts. This is achieved by applying the concept of constraint parametric models.

The following section gives an introduction to the types of manufacturing workcells that are considered. Afterwards, the concept for defining and accessing constraint parametric models will be presented. Then the problem of how to efficiently represent complex parametric models is dealt with. The last section presents the techniques that have been realized in order to evaluate and optimize layouts of complex manufacturing cells.

9.2 Types of workcells to be considered

Roughly speaking, two types of manufacturing workcells can be identified. The first type refers to cells that are a more or less loose composition of components. The number of functional dependencies is low. The only integrative element is that these components share the same location.

For the second type of manufacturing cells a close integration of components is characteristic. There are also many functional dependencies among them to be considered. The selection and assembly of components is very much restricted. This basic fact is of great importance for the modelling problem.

The following two examples outline this more deeply. Fig. 9.2 shows a KUKA IR161 in a flexible manufacturing cell for loading and unloading of machine tools.

Fig. 9.2   KUKA IR161 manufacturing cell – a loose component composition.

This is an example of a system with an extremely loose component composition. The components – arm extension of the robot, type of foundation, peripheral devices – can be selected by considering only the amount of valid values and the component position without regarding additional constraints.

An example for a closely integrated system is shown in Fig. 9.3. A KUKA gantry robot system is designed for a spot welding task.

The main components of this system are the rack, the bridges and the arm. It is obvious that a free selection of the components is not possible. There are heavy restrictions to be obeyed.

The characteristics of the different types of workcells which can be understood from these two examples have considerable influence on the type of models that have to be used. This will be outlined in the next section.

9.3 The modelling problem

The previous section has shown that particularly complex and closely integrated manufacturing cells require computer aided layout optimization. A very important problem is to have appropriate models available in order to represent these big systems and to allow access during the optimization process.

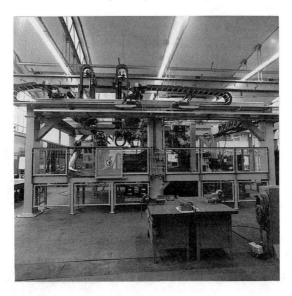

Fig. 9.3   KUKA IR400 manufacturing cell - a closely integrated system.

The concept of constraint parametric models is suited to meet these demands. It is a widely used concept even in other areas of computer application like VLSI design or software development.

Fig. 9.4 shows the representation scheme of the parametric model. It is composed of three parts. The nonvariable part of the model is represented by a set of structural elements that are fixed. The application of these elements is controlled by setting parameter values. The elements can be accessed by using symbolic identifiers within the parameter processing functions. The model structure elements are represented by convenient modelling systems.

The variable part is represented by functions, or laws. These parameter processing functions define how the invariant structure elements can be combined in order to get a complete model. The input to these functions are the parameter values. The parameter values are defined by their data type and the constraints they have to obey. Thus the generated model is an instance of the general parametric model applying a specific set of valid parameter values. A convenient way to realize the parameter processing functions is to use a procedural language. These procedures could be regarded as a list of adaption and assembly instructions for the structure elements.

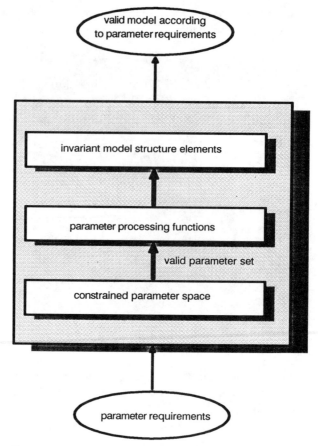

Fig. 9.4   Concept of representation of constraint parametric models.

The remaining problem is to represent the constraint parameter space. The following properties are typical for the parameters.

- The permissible values are of integer, real, enumeration or set type.
- Mostly they are of deterministic and not of stochastic nature.
- The structure of the parameters and their values change frequently.
- The parameters are heavily constrained.
- The constraints do not always represent analytical functions. Very often they have random nature.

Fig. 9.5 gives an example of a 3-dimensional parameter space.

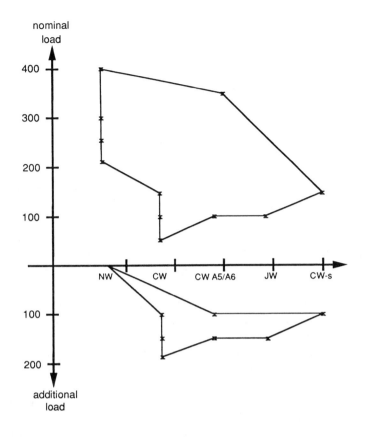

Fig. 9.5   Relational model.

wrist type	nominal load	additional load
no wrist (NW)	220	0
central wrist (CW)	60	80
central wrist - A5/A6 (CW A5/A6)	100	140
jointed wrist (JW)	100	140
central wrist -s (CW-s)	150	90
central wrist (CW)	60	140
⋮	⋮	⋮
⋮	⋮	⋮

The parameters are discrete. The boundary is arbitrarily shaped. Concerning practical applications, this concept has to be extended to an n-dimensional parameter space. A very general representation of constrained parameter space is to use relations and relational algebra. The compound parameter space representation can be evaluated by joining partial parameter spaces. Each single space is represented by a relation table.

The input to parametric models are the parameter requirements. Based on the internal representation of constraints that have to be considered, a valid parameter set is evaluated. This set of values is used in order to get an instantiated model that can be used for instance for simulation and optimization.

9.4 Representation of constrained parameter space

Initial applications showed that an efficient computer representation of the constrained parameter space is a decisive factor for a number of reasons:

- The dependencies between parameters are rather complex. In many cases it is not possible to represent them by using analytical functions.
- The number of variants of a product increases in the course of time. Concerning the model representation, this means that the size and the complexity of the parameter management increases as well, whereas the number of model structure elements and of parameter processing functions remains rather constant.
- There is frequent multi-user access to the parameter management area. So it is a big problem to maintain consistency.

From this follows that in many cases the constrained parameter space can be represented by using the relation concept only. For this relational data base management systems will be used. A valid parameter set can be extracted by using the 'select' operation.

Concerning practical application, some further boundary conditions have to be taken into account.

- For complex applications, more than thirty parameters are defined. However, the constraints will not be defined across the whole set of parameters but only for small groups of parameters (typically 2.4) which define a partial parameter space. The representation of the whole parameter space has to be evaluated by joining the partial parameter spaces.
- Normally, the partial parameter spaces affect different functional areas or subcomponents. Therefore the constraints are defined by different people. Thus in practice, they might be inconsistent.
- In order to get a valid compound parameter set, the partial relations have to be processed by 'join' and 'select' operations.

- The execution of a join operation on many relations with many records may consume some GByte of memory space. In order to avoid this the join and select operations have to be performed in an interleaved manner.
- A decisive factor is to process the partial relations in the right order. This is necessary in order to be able to process the partial relations sequentially with interleaving join and select operations.
- Expert knowledge is necessary in order to apply an appropriate processing order to the partial relations.
- The processing order also depends on the kind of parameter with which the user wants to start the evaluation.

In order to meet these boundary conditions, the following representation structure has been realised. It is shown by Fig. 9.6.

Two sections can be identified. The constrained partial parameter spaces are represented by relation data structures. These are managed by a standard relational data base management system.

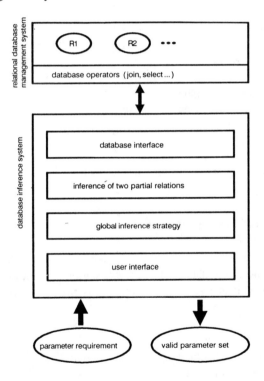

Fig. 9.6   Scheme of constrained parameter space representation.

The second component is the data base inference system. It is used to pick the different partial relations and fit them together. The result is a valid parameter value set across all parameters that meets the initial requirements.

This system basically is composed of four components. The data base interface provides access functions to the relational database management system, independent of a specific product. The next step is to evaluate two relations commonly and extract a valid partial solution. This is done by a join operation, succeeded by a record selection operation. This partial solution is the input for the selection operation of the next relation evaluation step.

The global inference strategy is provided by another module. It is up to it to decide about the next relation that has to be processed. Several levels of sophistication are possible. The simplest approach is to only allow a unique processing sequence. A more comfortable way is to offer alternative solution evaluation paths. When applying a declarative language, e.g. PROLOG, the implemented sequence of steps may be changed and improved more easily. The highest level of sophistication is reached by realising a totally dynamic access of all partial relations.

The user interface provides external access to the system, either for human operators or for external programs. The input are parameter requirements. These may be given in two ways:

- Only the values for some parameters are defined. The values of the others have to be evaluated, meeting some optimization criteria.
- Application-specific requirements are defined. These can be translated into parameter values directly.

The requirement input is supplied by the 'component selection and layout planning' module.

9.5 Model-based evaluation and optimization

The objective is to improve and optimize the model representing the manufacturing system. Fig. 9.7 shows the data flow of the evaluation and optimization process.

It is an iterative process. The information is processed by three function modules:

- evaluation and optimization;
- representation of the constrained parametric model;
- simulation of the instantiated model.

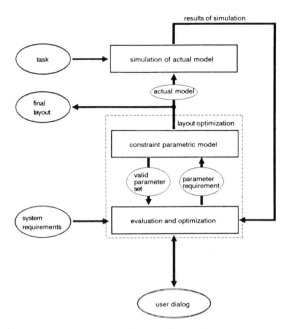

Fig. 9.7   Data flow during evaluation and optimization process.

The default layout information and the optimization criteria are provided by the layout planning module. The requirements for the design variables (parameters) of the model have to be evaluated next. This is done by processing the output data of the simulation system that represents the previous iteration step. During this process of defining the parameters for the next simulation step, the parameter constraints are evaluated.

The default layout information and the optimization criteria are provided by the 'layout planning' module. The requirements for the design variables (parameters) of the model have to be evaluated next. This is done by processing the output data of the simulation system that represents the previous iteration step. During this process of defining the parameters for the next simulation step, the parameter constraints are evaluated.

The next module is the constraint parametric model representation. This module serves two purposes. First it is used in order to get a parameter set that meets the constraints and that is supposed to improve the system properties. Then an actual system model is generated and supplied. This model is instantiated with the actual parameter set.

75

The instantiated system model is now used in order to simulate the task. The task itself is provided by the 'robot execution planning' module. During simulation, the relevant system criteria (objective functions) are evaluated. This is fed back again to the 'evaluation and optimization' module. The iterative process is continued until the optimization goal is achieved or the predefined maximum number of iteration steps is exceeded.

The question is how to realize the evaluation and optimization functions. Currently a totally automatic procedure is not possible. So the evaluation and optimization process has to be performed either by using fully interactive dialogue with a human operator, or by assisting him with specific functions.

The dialogue is supported by an appropriate representation of the information, gained during access of the constrained parametric model and during simulation. The information is provided in two ways.

- After occurrence the information is displayed on the screen immediately.
- The information is prepared internally and stored in a protocol file.

Using the application-specific functions, a semi-automatic optimization process can be realized. The realization of these user- and application-specific functions is supported by a user command language. This language is embedded into the user interface of the whole system. The language offers a wide range of features to calculate mathematical expressions, to control program flow, to process terminal input and output and to handle errors and exceptions.

The control of the simulator module and parametric model and the processing of its output data is fully integrated. So the command language can be used in order to control the simulation process and to realise optimization algorithms in a limited parameter space.

9.6 Summary

Layout optimization for closely integrated systems can only be executed by using constraint parametric models which are represented by relational data base management systems and expert knowledge. As the result of the layout optimization, the final specification of the layout components and their integration into workcell is finished. The design variables are defined by the parameters and the objective functions can be executed.

Chapter 10

Conclusion to systems planning

K. Tierney
University College Galway, Ireland

In general, robots are to be found in the three main branches of industry viz. the automotive, electric/electronic and mechanical engineering industries. Within these, the application areas can be divided into four: handling, machining, assembly and testing. However, it is clear that although robots are being employed in a variety of tasks and industries, they have not had as great an impact as might have been expected. A gap exists between the existing technology i.e. the large variety of industrial robots in the marketplace, and the number of applications of that technology in industry.

The lack of a *systematic planning procedure* for implementation of robot-based systems has been identified as a major obstacle in realizing the potential of robots in industry. It is an area which has received very little attention and one which is becoming increasingly important as manufacturing/assembly systems become more complex. Planning aids already exist for the application of industrial robots to simple tasks, such as the loading and unloading of machines. However, the planning of assembly is different to other areas of industrial application, for two main reasons:

- Generally, in the assembly process, several different parts have to be handled, in contrast to one or two workpieces in other applications.
- Assembly tasks frequently involve a relatively large volume of work, which leads to a sequence of assembly steps and to a sequence of assembly work stations.

These characteristics, together with a lack of knowledge on the assembly process itself and on flexible assembly systems design, highlight the need for a clear procedure to support the design process.

In view of this, the focus of this Section II has been on the presentation of a seven-phase planning procedure for the design of flexible assembly systems and on the development of methods and software tools to support each of these planning stages in an integrated environment. The planning procedure, which incorporates a top-down approach from rough planning through to detailed fine-tuning of the design, involves the following phases:

- analysis;
- outline;
- documentation;
- design;
- evaluation;
- detailing;
- installation.

Analysis involves the examination of the existing manufacturing/assembly system in order to establish the feasibility of automation. The interaction of personnel, means of production and workpieces is examined. Information is drawn from manufacturing programmes, engineering drawings, parts lists and on-site studies. The results of this stage determine the planning stages which follow.

Outline refers to the precise changes which must be defined in order to transform a manual system into an automated one. Existing processes need to be rationalised, work methods must be selected, and the degree of required flexibility must be estimated.

Documentation generates a list of requirements, in quantitative terms where possible, or in clear qualitative terms. These are stored in a database.

Design is the central planning task and involves the generation of a number of feasible layouts and equipment options, which are then evaluated using technological criteria such as cycle times and resource utilisation levels.

Evaluation focusses on the evaluation of the technological solutions from an economic standpoint. The goal here is to obtain a rough economic estimation before spending additional time and money in further detailing of a solution.

Detailing seeks the detail necessary for pilot production setup. Special devices may need to be designed, or existing equipment or procedures may have to be modified. The output from this design stage should include the number and detailed design of workstations, task assignment, production volume capacity of workstations, physical layout, location and capacity of work-in-progress buffers and reliability requirements.

Finally 'installation' involves the careful examination and planning of the installation of the proposed system in order to minimize disruption to the existing manufacturing system (Chapter 5).

An important feature of this planning procedure is its iterative nature. Planning work takes place in an environment which is constantly changing. Depending on the changes which occur e.g. economic or technological developments, product modifications or changes in market demand, the final solution may be influenced. Thus, the planning system must be able to incorporate these changes in the decision-making process as well as facilitate the designer in anticipating future influences.

Each stage in the planning procedure has been realised by one or more software tools developed by different partners within ESPRIT Project No. 623. The integrated planning procedure described in Chapters 6 to 9 starts with the generation of an operation sequence plan and preselection of robots, feeders and grippers. This information is generated by the manufacturing engineer using an interactive spreadsheet supported by a rulesbase (Chapter 6).

Based on the assembly sequence and the preselected robots, grippers and feeders, individual devices are chosen and roughly arranged on the shop floor. In an iterative manner, this initial layout is improved and verified using material flow simulation to estimate cycle times and resource utilisations for individual layouts (Chapter 7).

In the next step, these data are structured in a manner suitable for off-line programming. The assembly sequence and layout are described in a semi-formal language containing more detailed geometrical, technological and control specific data (Chapter 8).

The planning phase is completed with a layout optimization tool which considers the locations of robots, tools and peripheral devices and the influence they have on important parameters from cycle times to the total cost of the work cell (Chapter 9).

Based on the description of the assembly task and the optimized layout resulting from the planning system, programming of the robot is performed. This is the subject of Section III of this book and it addresses both on-line and off-line programming of robots for assembly.

To integrate each stage in the planning system, a central information system has been designed with standardized interfaces to allow information exchange and storage.

Section III

Programming

Chapter 11

Introduction to programming methods

K. Hörmann
University of Karlsruhe, Germany

Robots and other numerically controlled machines are being used more and more to automate manufacturing processes. Unfortunately, the programming of these machines is very time-consuming and expensive. Hence, it is very important to provide the users of these machines with powerful and comfortable programming tools.

The purpose of a programming system is to generate a robot control program in a user-friendly way. Compared to the programming of other numerically controlled machines, robot programming has some peculiarities:

- Robot programming is concerned with the generation of very complex motions. These motions are very hard to imagine for human beings without the help of the robot itself or a graphical simulation system.
- The geometrical data of the real world (e.g. positions and orientations of the parts to be handled) have significant deviations from those data used in the model of the real world which is used for off-line programming. Hence, very often sensors are used to measure these geometrical quantities. These readings are the basis of a conditional program to compensate the deviations.
- Typically, the robot has to be synchronized with peripheral devices such as sensors, end-effectors, feeders, fixtures etc.

Obviously, this list implies considerable requirements for a robot programming system.

The robot programmer must be able to specify the spatial operations of the robot in a simple way. The robot programming techniques must be adapted to the way of thinking of the user to facilitate the description of spatial operations. The user wants to program the robot in a problem-oriented way without the need to measure coordinates in the workcell and to explicitly specify these coordinates within the program. Furthermore, the system has to provide methods for real-time programming. The system has to support interrupt handling and the interaction between the robot controller and peripheral machines. To ensure that sensor data can be processed, means are necessary to program loops and conditional branches.

Summing up, a robot programming method must be able to do the following:

- be simple enough to be easily learned and handled;
- be problem oriented: the programming method must be adapted to the imagination of the user and to the particular application;
- support off-line robot programming in order to avoid using the expensive robot workcell for programming.

Today's robot programming methods fulfill only some of these requirements. The historical development of robots, starting with simple pick-and-place devices and leading to the present sophisticated devices, has been accompanied by the concurrent development of programming methods with ever-increasing power and complexity. The first robots were programmed with plug boards. This method was later improved by on-line programming methods.

The term 'on-line programming' is used if the programming process directly involves the robot itself. The main activity of on-line programming is the definition of trajectories. Usually trajectories are defined by specifying points (i.e. positions and orientations) with respect to the end-effector. The type of interpolation determines how the different limbs of the robot are moved and synchronized and how long this motion may last. The programmer may specify several types of interpolation together with such parameters as time, velocity and acceleration.

To specify the points of a trajectory, the robot is guided to the desired points and the corresponding joint values are recorded. The most common ways to specify these points are by specifying the motions on a teach-pendant ('teach-in programming') and by moving the robot via a master-slave linkage ('master-slave programming'). Sometimes the robot may also be moved manually while the trajectory is recorded as a series of closely spaced points. The latter method is especially useful for spray-painting.

Further improvements of this method were achieved by the incorporation of velocity, time, program branching, and many other special functions. On-line programming is still the most widely used robot programming method; even so, it is awkward for more complex problems. These methods are already well-known and hence are not treated in more detail in this chapter.

Increasingly off-line programming methods are used, especially for complex applications. Off-line programming involves the construction of a program text which is translated without the actual presence of the robot and is then afterwards translated or interpreted. A problem with this textual programming method is how to specify the movement points without the robot. Therefore, almost all textual programming systems are provided with a teach-in method, too.

In general, robot motions must be able to perform two tasks: reaching a programmed target position in order to position a workpiece or a tool or moving along a programmed curve with defined velocities. A planning module has been developed within the ESPRIT 623 project to generate executable trajectories which can be used as input by compilers or other program generation modules. This module is described in Chapter 12.

In textual programming, the operating sequence of the work trajectory of the robot and effector is written with textual instructions. During the initial development phase, such languages were obtained by extending existing programming languages with robot-specific instructions. The special robot programming languages were developed later. The planning modules developed in ESPRIT 623 generate an intermediate code which is called 'Explicit Solution Representation' (ESR). This code can then be translated into different languages for different controllers. This program generation module is described in Chapter 13.

To further simplify off-line programming and to shorten the time needed for program development, computer graphic tools are applied. Computer graphics can be used, for example, to simulate the effects of an off-line written program on a graphical screen. Still more user-friendly are graphical interactive programming systems. These systems allow one to specify motions in an interactive way and to immediately observe the effect of a command. Chapter 14 describes the Process Execution Simulation System provided by ESPRIT 623. It allows the testing of robot programs and the verification of whether the manipulation tasks will succeed or not. When executing off-line generated programs, often problems occur which are due to the accuracy of the model of the work cell and the positioning accuracy of the robot itself. Chapter 15 discusses solutions to these problems.

An even more advanced programming technique is the task-level programming method which is also called implicit programming. The term 'implicit' refers to the fact that these systems do not specify the robot operations in an 'explicit' way, as is the case with the methods already mentioned. Instead, the specification is given implicitly by defining the goal of the operations rather than the operations themselves, hence, the system automatically plans the action sequence needed to achieve this goal. A prototype of such a task-level planning system for assembly tasks has been developed within ESPRIT 623. It is explained in Chapter 16.

During the actual execution of robot programs, unforeseen situations may arise which can lead to a standstill of the workcell. However, automatic error recovery procedures can be applied so that the operation of the workcell can be resumed without operator interference. Work in automatic error recovery is presented in Chapter 17. Finally Chapter 18 summarizes and concludes Section III.

Chapter 12

Robot motion execution planning

G. Stark, P. Weigele
KUKA Schweißanlagen & Roboter GmbH, Germany

12.1 Determining factors of motion planning

The motion which a robot has to perform serves two main purposes. The robot has to achieve a programmed target position in order to position a workpiece or a tool. Secondly, the robot has to track a tool (or a workpiece) along a programmed curve with a defined velocity profile. This overall task is supplied by the 'process planning' module. The results are executable motion paths that directly serve as input for compilers or program generation modules. The objective of motion execution planning is to plan motion paths with respect to all influencing factors. Figure 12.1 shows a diagram of the determining factors. They will be discussed in detail.

12.1.1 Robot

The robot is the main component which has to be evaluated. The shape has to be considered concerning collision with obstacles in its surroundings and also concerning collision with the mounted tool.

The kinematic behaviour of the robot is determined by two facts. The kinematic structure (kinematic chain) is made up by the arrangement of joints (prismatic or revolute) and links. This is crucial to the working envelope and the quality of motion path execution. The second item are the motion capabilities. They are determined by the following conditions:

- range of each joint;
- velocity and acceleration /deceleration of each joint;

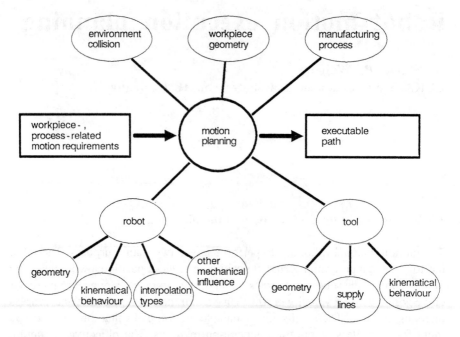

Fig. 12.1   Determining factors for motion planning.

- different modes for joint-, tool-center-point and orientation-interpolation;
- synchronization mechanisms;
- feedback of the joint level motion state to the cartesian interpolation process.

Last but not least, the various mechanical aspects of the robot have to be considered. Concerning high speed motion, the dynamic behaviour is important, since the contouring error is affected by this.

Concerning accuracy with low speed motion, the static forces/torques and their influence on flexible components are important. The following effects are relevant:

- static force/torques versus flexibility of gear and backlash;
- flexibility of robot arm;
- influence of payload;
- consideration of actual kinematic deviations (longitude, angle, structure).

12.1.2 Applied tool

The tool affects the execution planning in three ways. The shape, represented by geometry, has to be considered, in order to avoid collision. The position and orientation of the tool related to the flange determines the effective working space. This is extremely important when using multi-purpose tools, or tools together with changer. The kinematic behaviour of moveable tools like grippers or welding torch with additional joint has to be considered as well. A very important matter of consideration is the influence that the supply lines exercise. They restrict the effective working space of the robot and may cause collisions. The dynamic behaviour may negatively affect the execution quality.

12.1.3 Workpiece geometry

The shape of the workpiece, represented by 3-D geometry, is relevant in two respects. First, it defines the target position that has to be accessed or the curves that have to be tracked. The geometry of these elements has to be known very exactly. Further on, the shape of the workpiece may restrict the working envelope of the robot in the same way as the obstacles in the surroundings.

12.1.4 Manufacturing process information

The way the manufacturing process affects the motion is manifold:

- The movement of the tool center point has to be subject to well-defined velocity profiles. If the programmed profile cannot be maintained, the manufacturing process has to be controlled by signals that are derived from the actual speed.
- The orientation has to meet very specific process requirements. Very often, a constant angle between motion vector and main tool axis has to be ensured.
- The quality (precision) requirements are often very high. To achieve this, sections of the working space with reduced accuracy should be avoided. For instance, this applies to arm configurations where several axes are parallel or for the execution trajectories where joints change the sense of motion very rapidly.
- If kinematic systems are redundant, the additional degrees of freedom have to be used in order to increase performance and quality.
- If several motion robot alternatives exist, the best from the point of view of robot execution has to be selected.

12.1.5 Environmental information

The environment is made up of peripheral components or parts of buildings, restricting the effective working envelope of the robot. These obstacles are static or dynamic. The motion paths that have to be planned have to meet two requirements: collisions have to be avoided, and the trajectories, in order to pass around objects, have to be smooth. This is an important requirement for achieving high quality execution.

All the determining factors described above have to be processed. The next chapter outlines the topology of kinematic systems. Then the data representation of motion paths and the application of kinematic models will be discussed. Finally, the tools to support robot execution planning are presented.

12.2. Topology of kinematic systems

The topology of kinematic systems has to be understood in order to do proper execution planning. The problem gets rather complex if several kinematic units are involved. The base of a robot and the base of the motion path as well can be timed-variant when bound to a further kinematic system. The number of applications with complex kinematic systems is increasing. Figure 12-2 gives an example of a parallel arrangement of two kinematic units. A robot has to execute motion paths, bound to the platform of a 2-unit axis turntable. Robot and turntable share the same fixed base coordinate system.

In general, the topology can be described using a graph. The nodes represent cartesian coordinate systems, the edges represent the relationship between them. Basically, there are three different relationships to be considered:

- Fixed geometric reference: The base frame of a motion path or an execution unit is fixed to another reference frame.
- Kinematic chain: The endeffector frame (e.g. toolframe) is related to a base frame, applying n degrees of freedom.
- Execution of a motion path: An endeffector, represented by a frame (tool frame), has to execute a motion path that is also represented by an array of frames.

The representation and use of the motion paths and kinematic systems will be outlined in the following chapters.

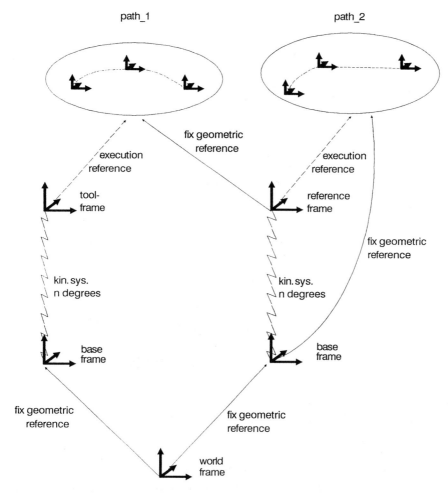

Fig. 12.2   Arrangement of kinematic structures.

12.3   Representation of motion path

An important issue is the matter of how to represent motion information. This can be done by either using data structures or language commands. The use of data structures is preferred. They offer a better way to structure the information that makes up a motion path and to do processing very homogeneously.

Figure 12.3 depicts a universal structure of a path data set. Two or even more kinematic systems are controlled by a common path data set. This is necessary in order to enable a synchronous path motion execution.

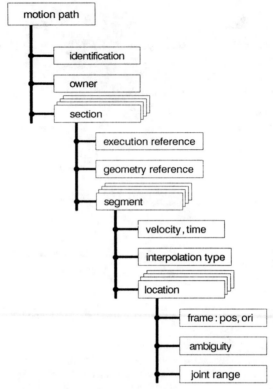

Fig. 12.3   Data representation of motion path.

The motion path has a unique identifier and is related to an owner program. The motion information for each kinematic subsystem is represented by a subpath called section. During execution, the different sections are processed commonly in order to achieve synchronous execution. A section is composed of 3 items:

- reference to an execution unit, e.g. a robot;
- definition of reference coordinate system (fixed or variable);
- list of motion segments.

Each segment holds a frame that defines its target location, the interpolation type to be applied and velocity control information. In order to get a unique solution, additional information for selecting the appropriate arm ambiguity and joint range state has to be added.

12.4. Representation and access to kinematic system

Simulation of motion paths is an important technique for robot execution planning. The planning functions have to have access to a simulation model of the system that has to be planned. Figure 12.4 shows the representation of the data structure of a universal kinematic model. The total system is composed of several partial systems that are to be integrated. Principally, they may be arranged in two ways:

- Sequential Arrangement: The endeffector frame of the first system is fixed to the base frame of the second.
- Parallel Arrangement: Either both base and/or both endeffector frames are tied together.

The integration part holds information about the number of partial systems and the type of arrangement.

The partial kinematic models have two main sections: the definition of the kinematic structure and the definition of the motion control data. The items that define the kinematic structure are listed below:

- Type of kinematic chain to be processed: This is needed to select the appropriate algorithm for the inverse kinematics problem.
- Joint level restrictions: These are limitations for angle range, max. velocity, max. acceleration/deceleration.
- Tree structure representation: It defines type and arrangement of the different joints to make up the total structure.
- Joint link bodies: These are identifiers of the geometry model of the bodies that make up the links between joints.
- Tool definition: This is comprised of tool transformation and the tool body.

The second data section applies to the definition of the motion control data. This is done by the following items:

- Joint level motion profiles: These are parameters to control joint level motion.
- Cartesian motion profiles: These are parameters to control cartesian level motion.
- Auxiliary time constants: These are parameters representing further time-relevant effects that have to be considered, e.g. interpretation delay time.

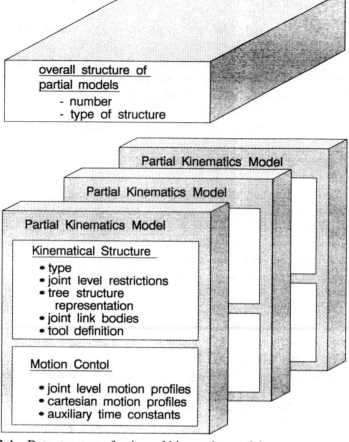

overall structure of partial models
- number
- type of structure

Partial Kinematics Model

Partial Kinematics Model

Partial Kinematics Model

Kinematical Structure
- type
- joint level restrictions
- tree structure representation
- joint link bodies
- tool definition

Motion Contol
- joint level motion profiles
- cartesian motion profiles
- auxiliary time constants

Fig. 12.4   Data structure of universal kinematics model.

12.5   Tools to support motion planning

The tools to support motion programming have direct access to the program interpreter and the simulation model. Three kinds of tools can be distinguished:

- tools for programming;
- tools for doing analysis;
- tools for calculation and optimization.

All tools are integrated into a homogeneous operating environment. Figure 12.5 gives an overview.

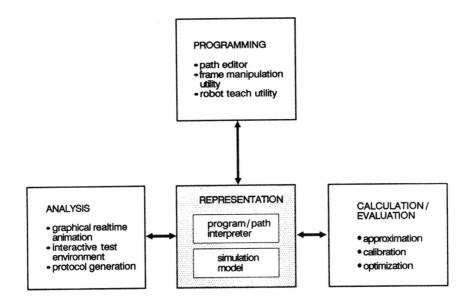

Fig.12.5   Tools supporting motion planning.

12.5.1   Programming

The motion that a robot or other device has to perform is defined on two levels. A path dataset defines all relevant information, excluding geometry. The geometry information is represented within the path dataset only by symbols. The information itself is maintained on a second level, which is the geometry modeller of the simulation system. A 'path dataset editor' gives support to efficiently define and modify the contents of the path datasets. It is a formular editor with automatic syntax and type checking.

The geometry information is represented by frame bodies. The 'frame manipulation utility' offers a large variety of functions in order to define, modify, and calculate frames.

- Frames are totally integrated into the geometry modeller. Instances of original frames can be generated and grouped into assemblies together with other geometry elements. This offers a high degree of power in order to structure frames and perform compound operations.

95

- There are many functions in order to manipulate frames, such as rotation, translation, mirroring, and arbitrary locating in space. The orientation of frames can be set according to optimization functions.
- Frames can be generated and located to workpiece entities such as points or intersections of curves automatically.

The 'robot teach utility' is used in order to locate frames related to the current position of the endeffector of a robot or another kinematic system. Powerful functions are needed in order to control and define the endeffector position interactively:

- The endeffector can be controlled, related to coordinate systems: joint, base, tool.
- The endeffector motion can even be related to variant coordinate systems, defined by the current state of another endeffector. The relation may be modified dynamically.
- Two kinematic systems, e.g. two robots, can be controlled synchronously.

12.5.2   Analysis

Analysis is an important planning function. Three kinds of functions are available:

- Graphical Real Time Animation: The use of visual aids is an important feature of interactive systems. This is especially true for the processing of geometry information combined with motion control. The execution planning system, therefore, is fitted with real time animation capability. This is achieved by applying graphics subsystems with powerful 3-D graphics processors. Real time animation gives a much better impression of dynamic effects. Two different viewing positions are possible: fixed position in space or position within endeffector coordinate system.
- Interactive Test Environment: The interactive test environment is used to test and optimize the execution of motion paths. It offers single-step mode with arbitrary step width and sense. The actual motion state is evaluated and displayed concurrently:
  - current position related to limit switches
  - current velocity
  - current acceleration.
- Protocol Generation: The interactive test mode can be enhanced by protocol generation functions. Thus, a more specific evaluation of the total motion path is enabled.

12.5.3 Calculation

The objective is to systematically define and improve the path datasets. This especially applies to the geometry information represented by frames. Three different kinds are to be distinguished: approximation, calibration, and optimization.

Most robot languages represent motion trajectories by a set of segments of only two cartesian interpolation types, linear or circular. A problem arises as soon as free sculpture surfaces have to be tracked. Programming can be considerably facilitated by the use of approximation algorithms. They automatically generate a set of segments of the allowable type that approximates the ideal curve.

Calibration procedures are necessary in order to adapt the offline-defined motion programs to the real situation on the shopfloor. This is done by generating correction transformations that are applied either individually to single frames or to a whole set of frames.

Optimization algorithms are also of great importance. They are mainly used in order to accomplish the following:

- generate smooth trajectories;
- achieve minimal execution time without exceeding the joint motion limitation;
- minimize contour error.

Optimization algorithms are realized by autonomous programs that have access to the path data set and frame representation utilities.

12.6 Conclusion

The objective of this module described in this chapter was to generate executable motion paths or trajectories which can now serve as input for compilers or program generation modules. Robot program generation is the subject of the next chapter.

Chapter 13

Robot program generation

W. Jakob
PSI GmbH, Germany

An important task for off-line robot programming is the connection of the programming system to a great variety of different robot controllers. To achieve this goal, a more flexible code generation is required.

The implicit or explicit components generate a program in an internal format which is called 'Explicit Solution Representation' (ESR) because it is the solution of the previous planning processes on the level of explicit programming. The ESR contains all necessary information for robot movements, data objects, program flow control, input/output operations, etc. in a neutral tree structure based on the type values. A tree node consists of the following items which can hold textual values except for the net relations:

- Type: The type describes the type of instruction (or part of an instruction) and it is used for defining structure rules for the tree itself. This item is mandatory.
- Attributes: An attribute can hold detailed information for the instruction, e.g. the target of a motion command. The number of attributes is type specific and may be zero.
- Comment attribute: This is a special attribute, the comment. It is optional.
- Text: Some types like 'program' or 'procedure' require a lot of space for comments, especially for documentation; thus, they may have a text item which may be of any length.
- Net Relations: There may be net relations between nodes, e.g. there is a net relation between the declaration of a data object and its applications.

The tree can be stored in a linearized form within a text file. The text format was chosen in order to improve portability.

The 'ESR File' is the input for the Code Generator as shown in Fig. 13.1. The code-generating process is controlled by the 'Robot Descriptor File'. This file contains generation rules that deal with the peculiarities of the various targets codes as well as with limitations of a specific control, e.g. the number of outputs. Occurring errors are reported in the file 'Error Messages'. The absence of this file indicates an error-free code generation. There are two standard outputs: the ESL and the DOC file. ESL stands for Explicit Solution Language. With a tree structure like the ESL, the kernel of the corresponding language is defined. A great Pascal-like presentation was chosen. The ESL can be regarded as a readable printout of the ESR. For documentation purposes, an enhanced ESL format, the DOC file, can be generated. It may be regarded as a pretty-printout of the ESL.

Within ESPRIT project 623, two target codes have been implemented and another two have been prepared for implementation:

SCRL: Assembler-like programming language of the RCM control of Siemens. The code generator has been interfaced with a RCM-3 controller.

MCL: Simple structured programming language of the ROSI simulation system of the Universtiy of Karlsruhe.

BABS: High-level programming language of the rho-2 controller of Rober Bosch GmbH (in preparation).

IRDATA: Standardized code interface (VDI 2863 or DIN 66313); see also Chapter 33.

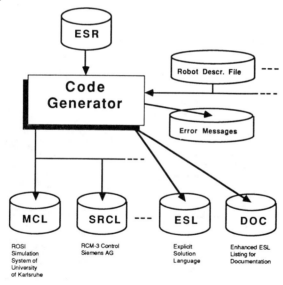

Fig. 13.1   System structure of the code generator.

Figure 13.2 gives an example of an ESR tree and the generated SRCL code for a FOR instruction. There are three code parts generated out of one ESR node:

- the initialization part, consisting of two load instructions (LAD);
- the compare-and-jump instructions (DEF...SPG);
- the finalization part, consisting of an increment instruction (ADD), a jump (SPG), and the definition of the exit label (DEF).

This example shows the benefit of the usage of the high-level ESR as a base for generating an assembler-like programming language. ESR is an intermediate code which can now be translated into different languages for different controllers. The next chapter deals with a graphical simulation tool for the testing and verification of robot programs.

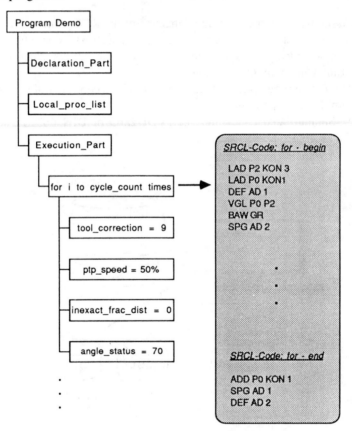

Fig. 13.2   Example of ESR tree and generated SRCL code.

Chapter 14

Process execution simulation

V. Katschinski, G. Schreck, R. Bernhardt
IPK Berlin, Germany

14.1 Introduction

The use of industrial robots in manufacturing systems requires computer aided planning tools for the manufacturing system planning as well as for robot programming. One tool is a simulation module within an off-line programming system. It enables the user to test and optimize application programs created off-line and can also be used for the planning and verification of workcell layouts. Both aspects can be realized within the simulation system. An important requirement for off-line programming systems is a test of the executability and practicability of the robot application programs created at the shop floor level to a maximum extent.

14.2 Overview and classification

The planning and programming of a robotized workcell requires a simulation system as an important part of the whole system. The simulator has to be in an adequate form to fulfill different demands. So it is necessary to distinguish the different forms of simulation.

On the one hand, a more communication-oriented simulation on a logical level is necessary for a cell, on the other hand a more motion-oriented simulation on the component (e. g. robot) level. Furthermore, some planning and programming features have to be given to the user for his disposal within the simulator.

14.2.1 Communication-oriented simulation

The simulation of the cell program execution in the distributed environment of a manufacturing cell is performed on a logical level. Modelling the cell environment in a task independent form requires the representation of the overall functionality of the cell. This functional representation has to include both static and dynamic properties of the cell. The two models are:

- the static model describing the layout, the components functionality and the relations between the components;
- the dynamic model describing the synchronization between the components (including the monitoring conditions) and the concurrency of the processes.

These two representations in connection with a trigger mechanism form the control model of the cell. The controller interprets the planned schedule of the cell and triggers the operations of the components according to the actual state and the enabled monitoring conditions. The actual motion simulation is performed according to the next chapter.

14.2.2 Motion-oriented simulation

A test of an application program for an active cell component like a robot will be done on the motion level. This practicability test contains testing of the defined trajectories related to positions, orientations, velocities and accelerations. Furthermore end-effector commands and interactions with peripherals have to be checked.

The simulator requires computer internal representation of all components of the workcell and their behavior according to predefined criteria. These representations are called simulation models and are structured in the following way:

- control models which describe the motion behavior of components;
- kinematic models which contain the frame relations of the different links;
- shape models which describe the graphical representation of components.

Figure 14.1 shows the functional connection of these models for the example of a robot motion simulation. The robot application program is loaded into the control model of the IR. This control model interprets the application program and supplies the kinematic model with the joint values of the links. Within the kinematic model the frames describing position and orientation of each robot part

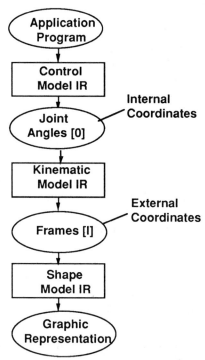

Fig. 14.1   The concept of motion execution simulation.

are calculated. The connection of these frames to the relevant shape models enables the visualization of the robot motion on a graphic system.

14.2.3   Simulation as a tool for planning and programming

The test of an application program by the simulator is one of the last steps in the whole planning and programming phase. The overall process is divided into different planning, programming and simulation steps where each step can be executed repeatedly. This iterative process requires process execution within the simulator based on the experience made in this phase. This programming feature can be for a single component (e. g. robot) and involves the derivation of intermediate points to ensure a collision free path of the robot. Another aspect of a planning feature is the rearrangement of cell components.

The cell oriented simulation system has to support the task-independent definition of the overall cell control model and planning and programming of manufacturing sequences. Cell control modelling, task sequence programming and

103

simulation are performed in subsequent steps. Cell control modelling requires as input the description of the cell configuration, its layout, and information about the components. This information is needed to build a functional model of the cell. This discrete model of the cell defines all possible states of each cell component, the synchronization between them and the possible parallel execution of the tasks. The generation of an executable manufacturing program for the cell requires as input a task specification and a description of the assembly parts. With this information the system plans the sequences of operations for each component of the cell. The resulting program has a hierarchical structure of tasks, subtasks and elementary operations.

The verification of planned cell program is performed in a dynamic process simulation, where the parallel execution of elementary operations, the logic control structure of the cell and the synchronization rules for each individual component of the cell are shown. This allows evaluation of the performance of the control model, detection of deadlocks and modification of the program.

14.3   Quality and validity of simulation results

14.3.1   Motion simulation/animation

The main task of a simulation system in an off-line programming system is the testing and verification of the application programs which were created off-line. Therefore, the simulator must allow the detection of errors in these programs. The difference between a simulation and an animation is that in an animation, the intended effects of the environment are programmed too. This means specific simulation commands are included within the application program to effect a change of the cell configuration, e. g. to grip a part during an assembly. In this case the programmer specifies the effects he wants to get as a result. Otherwise, if the simulator does such changing of states automatically, it is possible to check the application program and to detect errors.

14.3.2   Task   execution   simulation

The general process of off-line programming is shown in Fig.14.2. Out of a description of the production task, which forms the information basis for the programmer, application programs are generated. These application programs can be tested by a simulator. The simulator itself is configured to simulate the described production task and shows the effect the application program will have on the production environment. Also, for this configuration of the simulator the task description is the main information source.

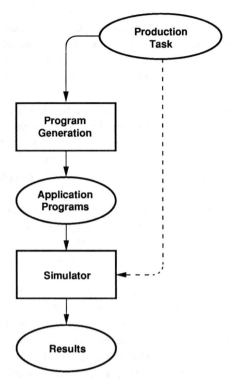

Fig. 14.2   A simulator in an off-line programming system.

However, as both the application program and the simulation are derived from the same task description they will fit each other very well even if they are incorrect due to faults of the task description. So what errors can be found by using a simulation to verify an application program?

The problem which presents itself is that task descriptions – depending on their level of detail – contain ambiguities, lack completeness and consistency and are often not entirely formal. Therefore, a formal basis for deriving application programs and their simulations is often not at hand. Consequently, the process of deriving application programs and their simulations has to be supported by humans who, on the one hand, are able to process ambiguous, incomplete and incorrect task descriptions but, on the other hand, lack accuracy. Even application programs and simulations which have been generated automatically do not feature the necessary level of completeness and therefore have to be checked and corrected.

One approach to the solution of the above-mentioned problem lies in the realization that application programs as well as their simulations are formalizations of a task description. Both are of course incomplete, but when the application programs are run in a simulator, inconsistencies between these formalizations become evident. Consequently, application programs and simulation have to be separated. This can be done in the following way:

The application program is the same program that will later be run in the real robot control. It influences the environment only by changing the output signals of the robot control but contains no information about the intended effects e.g. which part is gripped at which point of time during an assembly. The application program is a sequential algorithm and in this sense describes procedural aspects of the production task execution.

The introduced simulator program, on the other hand, describes declarative aspects of the production task. It defines which parts exist and which of them can be gripped under which conditions. This approach results in a test of procedural aspects of the simulation task against its declarative aspects.

In the following a more cell-oriented approach is outlined.

A cell task consists of a number of sub-tasks (abstract operators) which have a dependency relation given by the application. The dependency relations indicate the necessary order in which two tasks have to be carried out. The order constraints can be of a geometric nature like the assembly problem of stacking several parts upon each other. Also the availability of parts or production resources can impose constraints on the allowable order of the tasks.

A cell program is generated by transforming a global task description into a sequence of subtasks. This process is repeated until the subtasks are detailed enough to be directly expressed in the operating primitives of the cell components. These operating primitives are called elementary operations. The resulting task structure of a robot program and the corresponding elementary operations are graphically represented in Figure 14.3.

The task decomposition is done analogously for all active components of the cell.

In the simulation system, the planned cell program is tested and validated. Two different modes are possible for execution simulation: time driven and operation driven.

The time driven mode takes into account the duration time of the operations and triggers the process according to the planned due times. The latter mode requires feedback from the graphic simulation system where the actual action termination is reported and serves as input for the control model.

Both modes enable the simulation of the cell program flow and offer various analysis possibilities. One is the generation of interrupts in the execution simulation to verify the system state or to change a set of parameters and check the effect of the change after restart of the simulation. During simulation, the detection of deadlocks, due to unsolvable conflict situations or due to material flow reasons, is performed automatically.

The simulation system presents the status of each component with the actual active operation and a preview of the next operations to be performed. The trace flow of the execution is prompted to the user and is updated at each start of a new operation of one or more (i.e. parallel) components. The results achieved in this manner are then used to improve the quality of the former generated programs in automatic or user-controlled optimization loops.

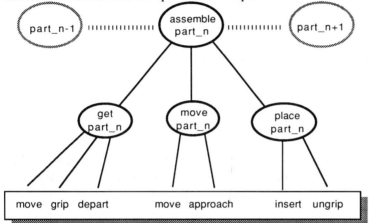

Fig. 14.3: Elementary operations and task knowledge for an assembly operation.

14.3.3 Test of application programs

Based on the task description the simulator has to detect errors in the application program relating to the practicability and completeness of the task execution. This includes a collision detection, a velocity and an acceleration check. Furthermore the interactions with peripherals have to be tested. The cycle time of the simulated application program is one result which allows an optimization in the direction of a synchronized overall manufacturing cell.

14.3.4 Differences of real and modelled components

The validity of a simulation depends on which way the simulation models fits reality.

The shape models of a workcell are based on the construction drawings or the related CAD-models. Differences between reality and shape model occur relating to cables and flexible pipes like air tubes for a tool exchanger. Such parts are normally not modelled and can cause problems in the task execution of the real workcell.

Kinematic behavior is based on the nominal kinematic description within the simulation system. Tolerances of the arrangement of cell components in the shop floor have to be considered.

The control model should have the exact behavior of the real controller. This control emulation requires the same algorithms used in the controller.

14.4 Requirements for system realization

14.4.1 Modelling of cell components

For the simulation system, computer-internal models are necessary which reproduce the components involved in the manufacturing task with regard to their motion behavior (control model), kinematics (kinematic model) and their shape (shape model).

The task of the control model is to interpret the application programs and to supervise the task execution. Therefore, the control model has to fulfill three functions:

- supervising the control structure;
- computing the joint values;
- communication with the environment.

All these functions of the controller model should work in the same way as the real controller.

The description of the kinematic should be realized in a universal manner to allow the representation of each kinematic chain.

For the graphical representation of the motion execution, each component within the work cell has to be modelled by its shape.

These shape models can be derived from the CAD-models. The accuracy of the models relating to the reality is an important factor with respect to the validity of collision detection.

To execute the cell task, the components have to cooperate. For this reason, a cell level control system is required to synchronize and monitor the processes in the cell. This cell controller needs, in addition to the given work plans, a task-independent representation of the target system, i.e. the real work cell. For this purpose, local device states and, as a combination of them, global cell states, have to be represented. In addition to these states, allowed transitions between the different states, i.e. the dynamic behavior of the system, have to be defined in the model.

To model the logical level of flexible manufacturing cells, the following requirements have to be fulfilled:

- Operating with various independent components, flexible systems are able to execute time parallel operations. Therefore, the model has to support the representation of the existing parallel processes running on different devices and the synchronization between them.
- The development of such complex systems is facilitated by dividing the system into subsystems. In this case, the specification method has to allow hierarchical structuring of the system into subcomponents. Further, the specification of the relations between the subcomponents has to be supported.
- Depending on the local capabilities of the single device control, i.e. the degree of autonomously executable operations, the level of detail represented in the model has to vary accordingly. The method can fulfill this requirement if it enables the specification of the components functionality on various levels of abstraction so that the specific model reflects only the necessary information.
- At all times the actual cell state has to be retrievable because checking of all possible states would require a large search space. Moreover, it would be reasonable to know the next reachable states to reduce the amount of conditions to be monitored before the next step.
- The processes in flexible manufacturing cells are event driven. In the model the discrete behavior has to be represented to simulate the processes. In addition to the simulation, which is the basis of improving the operation flow in the cell, the model has to ensure that the programs and the control system for the real plant can be applied.
- During simulation of FMS, various problems can arise due to unforeseen conflict situations and deadlocks. The model should be accessible to mathematical analysis and verification methods for deadlock and conflict detection in an early stage.
- In real processes, exceptions like loss of a part or positioning failures can occur. The exceptions first have to be detected and classified, and, subsequently, the planned recovery operations have to be executed. The selected method has to be accessible by dedicated exception handling procedures or systems (i.e. expert systems). Furthermore, it has to allow a restart from the interrupted system state without the need of restarting from the beginning.

- Finally, the specification method should be applicable in each project phase. The information representation has to be the communication link between the experts involved in the different phases.
- The definition of the functional model can be performed by a user in an interactive system modelling dialogue. The key to a complete and exact representation of the system lies in the definition of the possible interactions between the single components. As mentioned, these so-called relations are only dependent on the functionality of each cell component (i.e. which operations it is able to perform) and not on the task to be performed. The relations are strongly dependent on the layout. Thus, the layout description and also the component description with geometric data, kinematic data and technological data has to be known for the definition.
- One of the methods which meet the above-mentioned requirements are Petri nets with extensions proposed in recent publications. Because Petri nets are abstract models, it is possible to give an interpretation to the elements of the net assigning these to elements of the physical system. This assignment is more or less arbitrary and thus very important in the aim of having an exact representation of the system's behavior.

14.4.2 User interface

The user interface has to be realized in such a way that the operator can work with the system in an efficient way. This means the end user has at his disposal a menu oriented, interactive user interface which allows the user to work with application-specific functions. The visualization of the simulated workcell plays an important role in user acceptance. The graphical representation is the compensation for the real workcell (Fig. 14.4). It allows the user to see the motion behavior of, e. g., a robot in a normal way. Specific graphic features like translation, rotation and scaling support the interpretation of the scenario. The graphic interface should be realized by using standards (XWINDOWS, PHIGS) to guarantee a hardware independent solution.

14.4.3 Flexible configuration

A flexible configuration of the simulation system allows the simulation of different tasks. Therefore, a simulation language can be used to formulate the task execution requirements. This formulation describes declarative aspects of the production task. It defines which parts exist and which of them can be gripped under which conditions. This approach results in a test of procedural aspects of the simulation task against its declarative aspects. The advantage is that it not only visualizes the intention of the programmer but also the reaction of the robot's environment to the statements in the robot application program.

The initialization of the simulation system for the application should also be done by a data base. The restart of a simulation run should be possible from intermediate conditions to increase the efficiency of the tests.

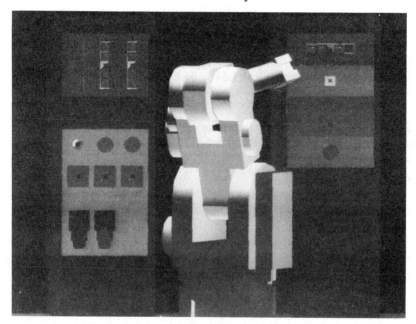

Fig. 14.4   Visualization of an A & R testbed scenario.

14.5   Concepts of realization

14.5.1   CAD system integrated simulation systems

One approach of realization is the extension of a CAD system by simulation features. Based on functions of the CAD system a new simulation-specific functionality can be built up, which allows the simulation of the task execution.

14.5.2   CAD system-independent systems

As an example for simulation systems independent from CAD systems, the robot simulation system ROSI developed at the University of Karlsruhe is outlined. ROSI enables the manufacturing engineer to perform the steps of a robot

111

application by means of graphical simulation, including robot selection, workcell layout, program development and test.

ROSI is based on a modular concept. The modelling component is composed of functions for the definition of geometry, kinematics, technology as well as functional parameters for manipulators and workpieces. For geometric modelling, the CAD-systems ROMULUS and EUCLID are used. The geometric data are transferred to ROSI as a part of the computer-internal model of robots, peripherals, sensors and workpieces. A specific subset of modelling functions supports the definition of a workcell layout. The user will be able to select the robots from the robot library of ROSI, which is stored in the data base of the system, together with all data describing machines, workpieces and the cell layout. Various emulators supply the system with software functions which allow the emulation of the robots, end-effectors, sensors, and peripherals. This includes robot independent trajectory planning, robot-dependent coordinate transformation, elementary motion functions of end-effectors and peripherals and sensor measuring routines.

The programming module allows the user to interactively develop robot programs using the internal motion command language of the system ROSI. The input of commands is facilitated by a hierarchical structured menu, which is presented on the screen, together with the graphical representation of the robot workcell. The specification of parameters can be performed by using submenus with graphical input features (picking of objects, adjusting of frames or robot joints).

A single parameterized command designates an action in the robot workcell, like a robot motion segment, a grasp operation or the evaluation of a sensor measurement. Step by step, a whole manufacturing program can be generated interactively, and will be immediately checked for constraints.

The graphic module allows the visualization of the robot workcell simulation using the capabilities of the employed graphic system.

14.5.3   Open systems structures

One simulation system which fulfills the requirements of flexible configuration for different tasks is a simulator developed at the IPK Berlin. This system has an open structure which is based on a modular concept.

This simulation concept which tests the procedural aspects of the statements in the application program to the declarative aspects of the behavior of the robot environment is outlined in Chapter 3.2. Based on this approach the following modular structure of the simulator´s kernel, shown in Figure 14.5, is realized.

The Communication Module provides several types of input and output buffers e. g. real, bool, integer, frame. It is the defined interface for the information interchange of the other kernel modules.

The Process Module includes the modelling of controls and physical aspects.

The Function Network supplies fundamental operations like AND, OR, DIST. It allows to define functions like the relinking predicates.

The Kinematic Model supplies parts and joints of several types e.g. ROT, TRANS, FIX. It contains the actual and possible kinematic trees.

The Graphic System is the defined interface to the graphic monitor. It initializes the graphic system, loads the graphical description of the parts and transmits the actual position and orientation of the parts to the graphic system.

The flexible configuration to applications is realized by defining a simulation language which can be interpreted. The Interpreter builds a Simulation Network which contains the complete set of data which are loaded into the Simulator's Kernel to execute a simulation task. The handling of time variant kinematic linkages can be defined with the simulation language, as well as the kinematic, signal exchange, controls and graphical representations. The link of different control models and external processes is foreseen. A control emulation delivered from a controller manufacturer can be linked to the simulation process via a defined interface. The same interface can be used for the monitoring of a task execution at a real workcell.

14.6   Outlook on further developments

14.6.1   Control model from the manufacturer

The control model should have the exact behavior of the real controller. This makes it necessary to use the exact algorithm of the real controller. The solution of the behavior for, e. g. a singular configuration of a robot is controller-specific. The controller manufacturer could deliver the control emulation software. This emulator could be linked via an interface with the simulator. The intention is to keep the controller model external with the advantage of being able to change it very easily. The simulator has to handle external processes to allow a simulation of different controllers within a workcell.

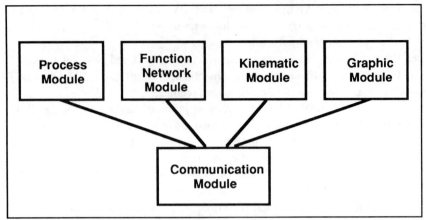

Fig. 14.5   Modular structure of the simulator's kernel.

14.6.2   Standard interfaces

An important role in the area of Computer Integrated Manufacturing is the standardization of interfaces to ensure the portability of applications.

Standards should be considered on every level of system realization. Based on this concept the existing standards e.g. for graphical interfaces like XWindows, PHIGS, etc., should be consequently used. New interfaces or model description, e. g. a kinematic description of robots developed in the system, have to lead to new standardization efforts with the goal of an easy exchange of models between different systems.

14.7   Conclusion

This chapter has described a simulation tool for testing the executability of robot programs. When executing such programs, problems often occur due to the accuracy of the model of the work cell and the positioning accuracy of the robot itself. The next chapter discusses solutions to these problems.

Chapter 15

Program execution

K. Schröer, R. Bernhardt
IPK Berlin, Germany

15.1 Introduction

An important goal in off-line programming concerns the creation of robot application programs which are executable to a maximum extent on the shop floor. This means that manual interactions to run application programs in the real robot (re-teaching) should be minimized. To reach this goal two important fields have to be considered. The first field concerns the communication link between the off-line area and the shop-floor. The second field deals with deviations between the nominal data used for layout planning and programming, and the real data of the work cell and the robot on the shop floor. Both fields are discussed in this chapter, which treats the following prerequisites and tools necessary for the execution of off-line generated application programs:

- Interface between off-line programming system and robot control;
- Protocol definition for data exchange;
- Code generator for the robot programming language;
- Robot calibration tools, which assure the required accuracy to allow the execution of off-line generated programs;
- Determination and compensation of environment and workpiece deviations.

15.2 Communication

Interface and protocol are presently determined and thereby hindered by the variety of interfaces and protocols used by different manufacturers. To overcome this problem, international standardization has been stressed since the beginning of the eighties. The publication of the Manufacturing Automation Protocol (MAP), version 3.0, has produced a future industrial local area network standard for shop floor communication. MAP is based on the ISO/OSI communication reference model (ISO 7498). Boardlevel and software products for MAP communication are now commercially available.The Manufacturing Message Specification (MMS) is a communication standard that fits into the Application Layer (Layer 7) of the OSI reference model. In October 1988 it became an international standard (ISO 9506). This version will remain unchanged within MAP 3.0 for at least six years. The purpose of MMS is to promote communication and interworking in a multivendor environment between shop floor applications.

MMS provides all services on the communication level which are necessary for the integration of robots into CIM systems: program transfer, program execution control, information and control of internal states and variables. Thus, on the communication level the obstacles for the introduction of robots into CIM systems will be overcome in the next few years.

15.3   Calibration

One main obstacle for the broader introduction of off-line programming techniques and the integration of robots into CIM systems is that a high absolute positioning accuracy of the robot is required for the execution of off-line generated programs. Otherwise a great amount of the time and cost which is reduced by off-line programming will be spent on the tedious, expensive and time-consuming task of re-teaching the robot.

The reason for this is not the lack of pose repeatability but the lack of absolute pose accuracy. The pose repeatability of robots available today is usually better than 1 mm. However, the absolute pose accuracy, i.e. precision with which the robot can reach a numerically given position and orientation relative to an external or work cell frame, is much worse. These errors are caused by differences between the dimensions of the articulated structure and those used in the robot control system model and by mechanical faults such as elasticity. These are systematic, deterministic errors which can be compensated if their parameters are known for each robot.

When teach-in is the robot programming method used, these deviations are negligible because the cartesian position and orientation of the taught pose are of no importance. The only thing which is interesting in this case is that the robot repeatedly reaches the taught pose with the required accuracy.

Only recently has an international standard (ISO 9283) been set, which defines performance criteria and related testing methods for industrial robots. What is here called 'absolute pose accuracy' is partly included in the ISO-defined criteria 'unidirectional pose accuracy', 'multi-directional pose accuracy variation' and 'distance accuracy'.

In order to close this gap between absolute pose accuracy and repeatability, robot calibration procedures have to be applied. This means, through TCP pose measurements at different robot positions, robot model parameters can be determined. This allows the improvement of absolute pose accuracy up to the basic deterministic accuracy of the system 'robot'. To realize this improvement, error compensation methods have to be applied which modify the off-line generated robot program by using the model parameter values for the robot which is to execute the task.

Procedures which only record deviations in TCP position and orientation in an inevitably small section of the six-dimensional robot workspace are not suitable, because compensation is only possible in the section from which the measurements have been taken. Therefore, the measurements must be repeated for each robot task and for each pose in which a high precision is required. On these grounds, calibration procedures have to be applied, which numerically determine the sources of the deviations and which do not need expensive experimental investigations or disassembly.

These calibration procedures must be able at least to identify numerically the exact geometric-kinematic parameters of the robot. The determination of elasticity and gear parameters should also be possible, because of the large number of robot types and application fields where elastic deformations, especially of the gears, cannot be ignored, if the desired precision is to be reached. One example of such a procedure for automatic robot calibration is described by Spur [1], Kirchhoff [2] and Duelen [3] and will shortly be presented here.

This procedure is based on a mathematical-physical robot model which includes all significant, deterministic sources of pose deviations, within which the geometric-kinematic model of robot motion is extended to include the effects of elastic deformations and gear parameters. In order to include elasticity, the reactionary forces and torques induced by the masses must be computed. (Fig. 15.1).

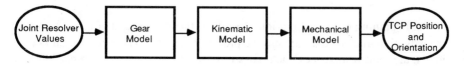

Fig. 15.1   Robot model.

For calibration purposes, the robot is considered as a stationary system in which the input values are the joint encoder values and the output values are the position and orientation of the TCP. Through numerical identification procedures the actual system model parameters are determined for the robot under investigation. The procedure allows the determination of all geometric parameters of the rigid body system such as zero position errors of the joints, link lengths, joint axis misalignments, transmission and coupling factors of the gears, gear and link elasticity, as well as gear eccentricity (i.e. periodic variations of transmission ratios).

The prerequisite hardware for automatic calibration is an interface between the process computer, the robot control and the measuring system.

An automatic theodolite system was chosen for the required measurements because it allows non-tactile position measurement in the entire robot workspace with the required precision (max. error smaller than 0.05 mm) (Fig. 15.2).

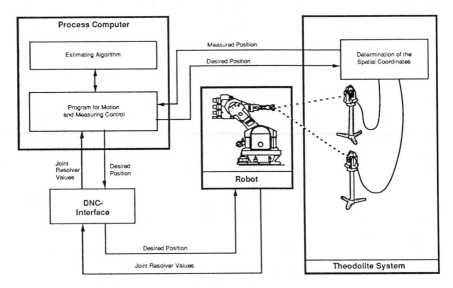

Fig. 15.2   Measuring process.

At the beginning of the calibration procedure, the known data of the robot under investigation must be entered into the process computer. These are the number of joints, the joint workspaces, the matrix of transmission and coupling coefficients, masses and centers of gravity of the links. The geometry of the kinematic structure (rotary or prismatic joint, position and orientation of the joint axis) can be entered as a simple polygon model.

Starting with this information, the internally used model is generated, its parameters are computed, and consistency and ease of modelling are checked. It is then automatically determined which robot poses shall be measured and used by the parameter identification procedure. The poses are generated with respect to parameter identifiability, observability and collision avoidance. The number of poses, which is chosen by the user, should be between 40 and 200.

During the measurements, the robot is commanded to the measuring positions; the actual joint resolver values of the target position are transmitted via the computer interface from the robot control to the process computer; using the known robot data, the approximate, actual TCP-position is computed and transmitted to the measuring system which then measures the exact TCP-position.

The calibration itself is now just the computation of those model parameter values which lead to an optimal fit between actually measured positions and those computed by the model. This is a non-linear least-squares problem, which is ill-conditioned and in which rank deficiencies can occur. Therefore, specially adapted numerical algorithms must be used. The result of the calibration is an individual parameter set for each robot which can, e.g. be delivered by the robot manufacturer.

The model parameter values identified in this manner can be verified by a procedure which determines the position deviations which remain when these values are used. To accomplish this, robot poses are taken which are also distributed over the entire robot workspace, but which may not have b... included in the identification procedure. By this verification procedure, the actually attainable absolute pose accuracy in the entire robot workspace is determined. The results of this calibration procedure [3] show that in this way the absolute pose accuracy can be improved nearly up to the limits (set by repeatability) of the system. Furthermore, by a thorough analysis of the numerical results, detailed hints can be obtained, by which constructive changes, even the repeatability and thus the basic deterministic accuracy of the robot, can be improved.

Because the available robot controls cannot make direct use of the calibration data, off-line error compensation methods must be applied in order to use the calibration results and to allow the execution of off-line generated application programs. On-line compensation would require too much computational power of the robot control, because the computation of reactionary forces and torques is necessary for the compensation of elastic deformations. If the algorithms of the robot control make use of explicitly invertible kinematic equations, then not even the set of geometric parameters identified by the calibration procedure can directly be used as a whole.

Off-line compensation methods allow the improvement of absolute pose accuracy using calibration results without any changes in the robot control algorithms and parameters. The prerequisite for this is that these algorithms and parameters are known. Then, using the powerful robot model f_M of the calibration

119

procedure and the model f_{St}, on which the control algorithms are based, compensation poses X_{in} are computed and transmitted to the robot control instead of the target pose X_{targ} which are taken from the off-line generated program (see Fig. 15.3; q : joint positions; f_R: function describing the real robot).

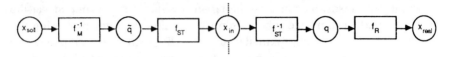

Fig. 15.3   Off-line error compensation.

One further application of calibrated robots is important when using off-line programming methods. It is based on the possibility to use the robot itself as measuring device.

After installation, the robot will be commanded to a small number of measuring points inside the work cell through which the frame of the work cell is defined. This frame is also the reference frame for the off-line generated application programs. Using distance sensors or other low-cost measuring equipment, the position of these measuring points relative to the robot base frame can be determined. Together with the known position of these measuring points relative to the work cell frame, the transformation between work cell frame and robot base frame will be computed.If this is done, the robot itself can be used as a measuring device to detect deviations between real work cell environment and the CAD models on which the off-line programming is based.

As a further requirement for calibration procedures, it must be possible to assure that the accuracy gained by calibration is kept even after repair. This implies the necessary possibility of recalibration, i.e. of determining a small subset of robot parameters which might have changed. The required measurements for this must be performed inside the work cell.

The whole procedure including model parameter identification and error compensation was successfully tested within the ESPRIT 623 project.

Another topic which should at least be hinted at here concerns changes and deviations of the work pieces which are caused by transport or machining of them. Recognition, control and compensation of these deviations need additional sensors and on-line measuring procedures.

15.4   Summary

The obstacles for the introduction of robots into CIM systems have been overcome to a certain extent. Based on international standardization activities, the obstacles on the communication level will be overcome within the next few years. Calibration procedures to improve robot accuracy are moving from research institutes into industrial application and thus will become standard remedies of the next decade. An alternative method of programming, that is, implicit or task-level programming, will now be discussed in the next two chapters.

References

[1]   Spur, G. and Schröer, K. (1989) Kalibrierung von Industrierobotern, in Vorschubantriebe in der Fertigungstechnik (eds. Pritschow, Spur, Weck), München, 1989.

[2]   Kirchhoff, U. Held, J and Schröer, K. (1988) Automatisierte Kalibrierung von Industrierobotern, in Komponenten für fortgeschrittene Roboter- und Handhabungssysteme, KfK-PFT 142, Kernforschungs-zentrum Karlsruhe

[3]   Duelen, G. and Schröer, K. (1990) Ergebnisse der Roboter-Kalibration, ZwF, Bd. 85, Nr. 2

Chapter 16

Task-level programming

K. Hörmann, G. Werling, B. Frommherz, J. Hornberger
University of Karlsruhe, Germany

A. Pezzinga, R. Gallerini
FIAR SPA, Italy

P. Bison, C. Mirolo, E. Pagello, L. Stocchiero
LADSEB CNR, Italy

16.1 Introduction

To solve an assembly task with the help of robots, many different problems must be solved: first, the necessary equipment and the layout for the assembly cell is determined by an assembler expert. Then, the robot programmer must commit himself to a sequence of single actions of the resources of the robot cell. He must also think about many geometrical problems, e.g. how to grip a workpiece in order to be able to perform a certain parts-mating operation, how to move a workpiece so that no collision occurs during motion, or how to use sensors to bring uncertainties under control.

A task-level programming system for robots must automatically find a sequence of actions in order to solve the specified assembly task. Closely related to this is the planning problem of Artificial Intelligence. Much research has been undertaken in AI, mainly concentrating on non-domain-specific planning or on block world problems. Aside from this AI-oriented research, there has not been much work on real-world robot planning. Early works include Taylor, Lozano-Perez [1] and Liebermann [2]. Taylor synthesized sensor-based programs of the robot programming language AL by the parameterizing of prototypical strategies (so-called procedure skeletons). Such skeletons contain a framework of motions, error checks, and computations for a particular type of task. The planner performs geometrical computations and error computations and decides which strategy to apply and how to parameterize this strategy. A similar approach based on procedure skeletons is taken in the LAMA-system. For AUTOPASS, the syntax and semantics of a task-level robot programming language has been defined. An emphasis of this research is an algorithm for collision-free path-planning for a robot.

The TWAIN-System proposal [3] includes modules for layout planning, fine motion planning, grasp planning, gross motion planning, and the selection of feeders and fixtures. The task specification is hierarchically decomposed into planning islands. Constraint propagation of symbolic expressions of planning variables is used to communicate between planning islands to find instantiations for the planning variables. If the planning variables cannot be instantiated, backtracking is used to modify the decisions made up to this point. The SHARP-system proposal consists of modules for grasp planning, fine motion planning, and gross motion planning. It also uses constraint propagation to coordinate the modules. HANDEY [4] is an implemented system for the planning of pick-and-place operations. It uses information provided by a 3D vision system. HANDEY plans both grasp, and if necessary, re-grasp operations.

Within ESPRIT Project 623, such a 1 system has been developed. Its subsequent planning phases are shown in Figure 16.1. The system is a tool for automatic off-line robot program generation for assembly tasks. It determines for each robot of a workcell the program which, when executed, solves a given assembly task, resolving problems of coordination and collision avoidance between the robots. These programs are generated starting from a geometrical description of the parts to be assembled and the way they have to be connected (Task Description), and the definition of the workcell that has to execute the assembly task (World Model). The system is composed of different modules which are briefly introduced in the following text.

The first step is to define the assembly task in terms of the spatial relationships among workpieces. For this purpose a graphical editor has been implemented (see Section 16.2) that allows these objects to be manipulated interactively.

Next it must be determined in which order the various parts have to be assembled, starting from a geometrical description of the parts and the way they have to be connected. This order is expressed by an 'augmented assembly graph'. Conventional assembly graphs do not state explicitly what operations can be executed in parallel, or what sub-assemblies can be built in solving the assembly task. This module (see Section 16.3) produces a precedence graph in which parallelism in operations is expressed in the creation of sub-assemblies in different areas of the workplan, and in the explicit definition of which operations can in principle be executed in parallel. This information is used to plan and coordinate the actions of the robots of the workcell in order to reduce the time needed for executing the overall assembly task.

The Scheduler (see Section 16.4) assigns assembly operations to the robots of the workcell using the ordering and the information about possible parallelism of operations determined by the Procedure Graph Generator. The goal is to produce a plan of operations for each robot of the workcell that is 'executable', meaning that it is actually executable by the robots and that it avoids collisions between the robot and the parts, or between two robots. Among the possible executable plans, the Scheduler tries to determine the one corresponding to the minimum overall assembly time. The Scheduler estimates the time needed for assembly operations to synchronize the actions of the various robots.

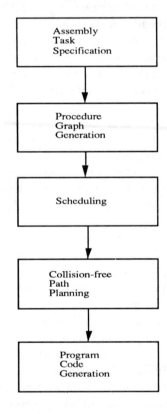

Fig. 16.1   Subsequent planning phases of the task-level programming system.

Each operation of a plan determined by the scheduler must be expanded into a set of executable actions. This requires collision-free path planning (see Sections 16.5 and 16.6) for the various types of robot motions such as graphic operations, fine motions, and transfer motions.

In the last step, an executable program must be generated. This code generation process is described in Chapter 13.

16.2 Specifying configurations of 3D-objects by a graphical definition of spatial relationships

Configurations of three-dimensional objects can be defined by sets of coordinates for each object which are related to a world coordinate frame. To calculate these coordinates it is necessary to know the exact size of the different objects. A more comfortable method is to specify the position of an object by defining a set of symbolic spatial relationships between object features (faces, edges, vertices). The position of the object is then calculated implicitly. However, the textual specification of spatial relationships forces the user to generate a name for each object feature. This problem can be solved by using a graphical system for the specification of relationships. In this section a system is presented which allows the graphical implicit specification of scenes.

16.2.1 Introduction

In many applications it is necessary to place a set of geometrical objects somewhere in a scene. The interactive planning of the layout of an assembly cell for example consists of placing robots, feeders, workpieces, etc., within a given area. In task-oriented robot programming it is necessary to specify the assembly task which enables the planning system to generate the corresponding robot program. One part of this specification contains pure geometrical information about the position and orientation of each part of the assembly. It is important that this information is represented independent of the robot type, since the robot which has to perform the task is selected in a separate planning phase.

In a three-dimensional space the position and orientation of an object can be described unambiguously by a set of six coordinates. This can be achieved by embedding a local coordinate frame in each workpiece and by defining one global reference frame anywhere in the space. The location of an object can now be defined by the displacement and the twisting of the local coordinate frame of the object with respect to the global reference frame. Normally, three coordinates are used for the description of the translation which refers to the main coordinate axes x, y, z [5]. The orientation of an object can be defined by three angles. Specifying

125

locations of objects by Cartesian coordinates is a common method which is also used in robotics to define the goal positions of the tool center point of a robot. Therefore, many of the modern robot programming languages provide language constructs for such specifications [6]. However, this approach has some drawbacks: To obtain the coordinates of a workpiece it is necessary to calculate the single values by using a technical drawing of the assembly. This is very tedious, since not all dimensions are directly given in the drawing. In addition, if the location of a symmetrical object is specified by six coordinates, it is mostly overspecified, since the rotational degree of freedom resulting from symmetry has to be arbitrarily defined.

A second approach for the specification of part locations which avoids these drawbacks is given by Popplestone et al. [7]. Instead of using coordinates, the goal position of an object to be moved is given by a set of symbolic spatial relationships between some features of this object and some features of other objects in the scene. Features of polyhedral objects are geometrical elements like object faces, edges, vertices, and auxiliary elements like axes of symmetry etc. From a set of spatial relationships a location of the object to be moved can be deduced where all specified relationships are fulfilled. Examples for spatial relationships are: two faces are in direct contact, or two edges are parallel and have a distance of 10 mm, etc. This is called an implicit specification, since the coordinates of the goal position of the part are not explicitly given by the user, but automatically deduced by the system. An object position can be completely specified by two or more spatial relationships between the object and other objects in the scene. In many cases such a list can be reduced to one strong relationship, if pairs of the given relations are repeatedly substituted by one 'stronger' relationship. 'Stronger' means that this relation compresses the information of two parent relations into one. The new object position can then be extracted from the resulting spatial relationship. However, if the specification is done textually, it is necessary to generate a name for each object feature to be identified. This can also be very tedious, since the number of object features may be very high even for simple objects. The solution of this problem can be achieved by combining the implicit specification method with an interactive graphical tool. Instead of referring to object features by name, they can be easily selected by a graphical input device like a tablet. The advantage of this kind of specification is that in most cases, it is not necessary to know the dimensions of the different parts involved. Instead of calculating coordinates of a single object, features are identified and spatial relationships between them are specified. It is important to mention at this point, that the method of implicit specification is not always easier than the explicit one. If, e.g., an object does not fulfill simple spatial relationships at its goal position, auxiliary geometrical elements must be generated to specify the desired location.

Therefore the assumption is made that a system for the specification of configurations should combine both the explicit and the implicit method.

The system described in the following sections is a part of the user interface for a robot action planning system called GRIPS (GRaphical Implicit Programming System) which has been developed at the University of Karlsruhe. The reduction of lists of relationships is performed by a term rewrite system which represents the kernel of the implicit positioning module. This subsystem will be described later in more detail.

16.2.2   User interface

The set of commands offered to the user can be divided into four subsets: scene modelling, positioning, utilities, and miscellaneous functions. The subset scene modelling includes all commands to get object models onto the graphical display and into the internal world model and to remove them if they are not needed any longer. 'Get scene' and 'Put scene' are commands to respectively store and restore whole scenes consisting of objects and their environment. This feature makes the modelled scene available to other modules of the system or to other systems. Moreover, it enables the user to terminate an incomplete session at any time and to resume it later. 'Insert object' is a basic command for modelling scenes. It inserts an object into the current scene and places it into the origin of the world model. From there the object can be moved to any position in the scene by using special positioning commands. An object can be removed from the scene by the command 'Delete object'. After a scene is composed, it can be manipulated by using commands of the subset positioning. These commands are divided into two further subsets containing operations for explicit and implicit positioning.

The more traditional way of moving objects within a world model is by using explicit positioning commands. This means that the user will numerically specify the coordinates of the goal position with respect to the world coordinate frame. For many applications it is useful that one can choose the coordinate system, to which these coordinates should refer. Therefore, the user can choose between 'absolute' and 'relative' interpretation of his specified coordinates. Absolute interpretation means that the given coordinates will refer to the basic coordinate frame of the scene. For a relative interpretation, the user can select either a local coordinate frame of an existing object or define a new suitable coordinate frame. This is very useful for operations like moving an object parallel to the surface of a Table or moving a gripper relative to an object to be manipulated. Coordinate values can be specified either by a textual input via keyboard or via control dials.

A second method to move objects on a graphical display is given by the subset for placing objects implicitly. It enables the user to specify a new object position by defining a set of symbolic spatial relationships between object features as described in the previous section. Possible object features are plane faces, edges and vertices. The feature type 'edge' includes the normal edges of the objects and additionally defined axes of rotation like symmetry axes of pins, holes, etc. Vertices are defined by the intersection of two or more edges. A new object position is specified by a sequence of one or more spatial relationships which the object should fulfill at the intended goal position. After every specified relationship a new object position is determined, where the moved object meets all previously defined spatial relationships. Since the object at the new position is shown on the display, the user is able to incrementally define the desired final position. This procedure is rather similar to the method of placing objects in the real world. When the moved object is at the intended position, the sequence of relationships must be concluded by a 'Close'. Then another object can be manipulated. Object features can easily be identified by selecting them on the display using graphical input devices, e.g. a tablet with a pen. An edge can be identified by selecting it with a graphical cursor, a face by selecting two of the bounding edges and a vertex by selecting two intersecting edges.

The system provides several utilities to support the positioning procedures. For symmetrical objects like cylindrical pins etc., one can often make good use of their axis of symmetry. Normally, these axes are not included in the graphical models and therefore cannot be used. The definition of auxiliary planes might also be very useful to obtain 'stronger' spatial relationships. The 'Create frame' function is necessary if relative explicit positioning commands are used. The user can define new coordinate frames anywhere in the scene and refer to them during object manipulations. The penetration check provided by the system enables the user to test whether the objects of the scene penetrate each other. This function is important, since it is very difficult to detect penetration of non-convex objects by inspecting them on the screen. The fourth subset contains functions for adding further, invisible information to the scene. Some applications like task oriented robot programming and computer vision can make good use of user-defined attributes, associated with objects or object features. The system allows the specification of attributes, such as the weight and kind of material of an object, the roughness and colour of an object face, etc. Moreover, it is possible to specify more general relations between objects or object features, e.g, 'object A is connected to object B' or 'object A lies on object B'. This information is useful for positioning functions or for other systems using the modelled scenes.

16.2.3 System architecture

The user interface consists of four main components. These are as follows:

- the menu manager;
- the input manager;
- the command interpreter;
- the term rewrite system.

The menu manager builds up the menu structure and handles the menu input. Since the menu definition is held on a special description file, the menu is very flexible. When a menu item is picked, the graphical pick information is read via the graphical interface. The corresponding command is encoded and submitted to the input manager.

The input manager is concerned with both textual and graphical input handling. During textual input, it analyzes the command and submits it to the command interpreter. During graphical input, the input manager leads the user to a correct input. The implemented input mechanism can be regarded as a finite automaton where transitions between different states are caused by edge picking.

The command interpreter mainly consists of three submodules which are concerned with scene modelling, positioning and miscellaneous functions as described above. The positioning module provides all functions necessary for the numerical and symbolical specification of part positions.

The term 'rewrite system', which processes the reduction of sets of spatial relationships, is part of the submodule for implicit positioning. This subsystem is explained in the following section.

16.2.4 Reducing sets of spatial relationships using a term rewrite system

The aim of the term rewrite system (TRS) is to deduce an unambiguously defined object position from a given list of spatial relationships. In order to specify such a list, it is necessary to provide a language. A word of the language is represented by a list of spatial relationships. One relationship consists of a symbol defining the kind of relationship, two specifications of object features of the two objects involved, and the intended distance between these features. Object features are planes (PLANE), edges (EDGE), and vertices (VERTEX). A plane is specified by its normal vector and one point, both given in Cartesian coordinates. An edge of an object is defined by a point and a direction. A vertex is represented by a three-dimensional vector.

The following list contains the different types of spatial relationships which can be specified between object features and explains what they mean:

FIX   No degrees of freedom exist between the two objects involved.

ROT   Only one rotational but no translational degree of freedom remains between the two objects involved.

LIN   Only one translational but no rotational degree of freedom remains between the two objects involved.

AGPP   This relationship can only be specified between two planes of two objects. It means that two planes are positioned such that the normal vectors of the planes are directed in opposite directions.

COPL   Like AGPP, however, the normal vectors of the planes have the same direction.

AGPE, AGEP   These symmetrical symbols are defined between a plane of one object and an edge of another object or vice versa. It means that the edge is perpendicular to the normal vector of the plane. There are both translational and rotational degrees of freedom remaining.

AGPV, AGVP   These symbols are defined between a plane of one object and a vertex of another object or vice versa. They can be used for simulating the relationship 'sphere against plane'. There are both translational and rotational degrees of freedom remaining.

AGEE   This symbol is defined between two edges of two objects. It means that the directions of two edges are parallel or antiparallel. There are both translational and rotational degrees of freedom remaining.

AGEV, AGVE   These symbols can be used to define a relation between an edge of one object and a vertex of another one or vice versa. It allows the simulation of the spatial relationship 'cylinder against sphere'. There are both translational and rotational degrees of freedom remaining.

AGVV   This symbol can be used to define a relation between two vertices of two objects. It allows the simulation of the spatial relationship 'sphere against sphere'. There are both translational and rotational degrees of freedom remaining.

The relations FIX, ROT, and LIN usually result from a rewriting step and cannot be specified by the user. These relations and the relations which contain a plane as an object feature are 'strong' relationships. Only few of them are required for achieving a completely specified object position. Most other relationships are used only for consistency checks.

The syntactic structure of the language in a BNF notation is as follows:

```
relationship_symbol   ::=   FIX | LIN | ROT | AGPP | COPL | AGPE | AGEP
                            |AGPV | AGVP | AGEV | AGVE | AGVV | AGEE
feature               ::=   PLANE   |   EDGE   |   VERTEX   |
                            COORDINATE_FRAME
```

spatial_relationship	::=	(symbol,distance,feature_moving, feature_fixed)
symbol	::=	relationship_symbol
distance	::=	REAL VALUE
feature_moving	::=	feature
feature_fixed	::=	feature
problem_specification	::=	(spatial_relationship)$^+$.

The following section describes the algorithm on which the TRS is based:

The TRS tries to reduce a set of spatial relationships to one single relationship containing the information of the whole set. Starting with a problem specification, the system repeatedly tries to apply reductions of the type D--> D' with D, D':= (spatial_relationship)$^+$, until a final state is achieved. The reduction denoted by "-->" consists of four types of rewriting rules (sr,sr1,sr2 are of type spatial_relationship):

- Rules of the form (sr1,sr2)--> sr mean that the single relationship sr satisfies both, the relationship sr1 and the relationship sr2;
- Rules of the form (sr1,sr2)--> sr_1,...,sr_n lead to several alternative solutions for one rewriting step. Each of the resulting spatial relationships sr_i with i=1,...,n contains the information of both parent relationships sr1 and sr2;
- Rules of the form (sr1,sr2)--> (sr1,sr2) indicate that there exists no solution for a given pair of spatial relationships, i.e., it cannot be substituted by a single relationship;
- Rules of the form (sr1,sr2)--> error mean that an inconsistency between two spatial relationships was detected.

After starting the algorithm with input S it will run until one of the following final states Fi is achieved:

- F_1 = {spatial_relationship I symbol=FIX}
  This means that a final result was achieved. This is equivalent to a completely specified object position without any remaining degrees of freedom. However, this does not mean that there exists only one final relationship of the type FIX. Rewriting rules of the second type may lead to several final results, each entirely satisfying the initial problem specification S.

- F_2 = {spatial_relationship I symbol Œ relationship_symbol \ {FIX} ∧
  sr_i.symbol = sr_j.symbol}
  This means that no final result was obtained. This state will be achieved if,

e.g., the object position is incompletely specified. The algorithm could reduce the set S to one or more spatial relationships of the same type.

- $F_3 = S$
  If the algorithm cannot achieve a single type of relationship (neither F_1 nor F_2), the resulting solution is identical to the problem specification S. This case indicates that S bears too few or too weak spatial relationships. To obtain a significant result, it is necessary to specify additional relationships. This state is denoted by no rewriting result.

- $F_4 = S$ with S.state = error
  This means that an inconsistency between two spatial relationships was detected.

For most applications it is necessary to achieve a final state of type F_1 i.e., all existing degrees of freedom of the objects involved must be removed. For some applications, however, it is adequate to attain a final state of type F_2. If a cylindrical object like a pin should be inserted into a cylindrical hole, it is sufficient to achieve a final relationship of type ROT. Any further specifications leading to a final state of type F_1 would result in an overspecification of the problem.

   The algorithm is started with the problem specification as input. It will run until one of the possible final states is achieved. As can be seen from the rewriting rules above, the algorithm operates on pairs of relationships taken from S. If a rule of the type 1 can be applied successfully, the two parent relationships sr1 and sr2 are removed from S, while the resulting relationship sr is added. The algorithm continues with the set $S' = S \setminus \{sr1, sr2\} \cup \{sr\}$. Applying a rule of type 2 will lead to a new set of specification sets $S'i = S \setminus \{sr1, sr2\} \cup \{sri\}$, i.e. one set for each resulting alternative relationship of the previous rewriting step. Each $S'i$ is separately processed by the algorithm and can produce a distinct final state. If only a type 3 rule can be applied to a pair of relationships from S, it is marked as 'being processed' but not removed from the set. If a type 4 rule can be applied, the algorithm stops processing the current set and returns to the top level. The termination of the algorithm is guaranteed by the rule types 1, 3 and 4. The operation will stop if either no further rule can be applied or all relationships from all sets are marked. An informal description of the algorithm is given in the following.

0    $Aj \in D$
1   I:=S {I is an auxiliary set, initially containing S}
2   R := { }

```
3      WHILE R ⊄ F_i AND not every alternative has been processed
       DO
4          { try next alternative}
5          WHILE |I| > 1 AND a rewriting step is applicable
           DO
6              choose (sr1,sr2) such that a rewriting step is applicable
7              (sr1,sr2) → result {execute rewriting step}
8              IF result = error
9                  THEN I.state:= error
10             ELSE IF result=(sr1,sr2)
11                 THEN mark sr1 and sr2 as being processed
12             ELSE IF result=sr_1,...,sr_n
13                 THEN (create alternatives
                       Aj :=(I \ {sr1,sr2} ∪ {sr_j};) I:=A_1)
14             ELSE {result=sr} I:=I \ {sr1,sr2}∪ {sr})
15             FI FI FI
16         OD
17         IF I.state ≠ error AND |I|=1 THEN R:=R ∪ I  FI
18         I:=A_j {next alternative}
19     OD
20  IF R={} THEN R.state := error  FI
```

F_i can be one of the four final states mentioned above. The used variables S, I, and R represent sets of spatial relationships, where S is the problem specification and R the resulting list. The variables sr, sr1, sr2, sr_1,..., sr_n represent single spatial relationships. For more detailed information see [8].

16.3   Automatic generation of precedence graphs

Precedence graphs are a convenient method to represent a valid order of assembly operations. First, this section shows how elementary precedence constraints can be deduced automatically from a CAD model of the assembly task and how they can be integrated in a precedence graph. Second, it discusses the overall structure of a planning system which operates on the basis of precedence constraints.

16.3.1   Precedence constraints and precedence graphs

Precedence graphs are a convenient method to represent the flow of assembly operations. The nodes of these graphs describe the various assembly operations which are necessary to execute a specified assembly task. The edges define the chronological order of the operations. They express the relation 'before', which means that one operation can only be performed if another operation is completed.

Precedence constraints may have various causes which result either from the task itself or from the resources like robot, gripper, etc. which were selected to execute the task. For the successful execution of an assembly task, several rules have to be considered according to physical laws. The following list specifies some of the most common rules which are given only by the assembly task:

- Rule 1: The parts of an assembly must not penetrate each other during the assembly operations. The example in Figure 16.2a shows two parts A and B in their goal position. Since the part B is 'external of' A, it is necessary to assemble part A first. For this precedence relation, the notation $\pi(\vec{A}, \vec{B})$ will be used which means: assemble part A, then assemble part B;
- Rule 2: Each intermediate assembly state must be stable i.e., every part must keep its position after it has been assembled. If, e.g., a part B has to be placed on top of part A (see Figure 16.2b), then part A must be assembled first. Again the resulting precedence constraint is $\pi(\vec{A}, \vec{B})$;
- Rule 3: The danger of dislocating already assembled parts during an assembly operation should be kept to a minimum. In the example shown in Figure 16.2c, two flat parts A and B are to be fastened by a pin C. If C is assembled last then B could become disarranged. One possible solution could be to assemble B before C The general rule is BC.

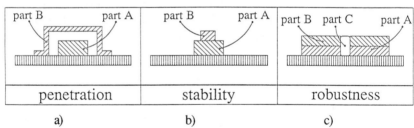

penetration	stability	robustness
a)	b)	c)

Fig. 16.2   Analysis criteria for assembly tasks.

This set of rules is certainly not complete. There may be several other criteria which must be observed in order to find a good assembly sequence. The list can be freely extended and the resulting precedence constraints can easily be used to find an overall plan. This will now be demonstrated by applying the above three rules to another example, and showing how the resulting sets of constraints can be integrated in one precedence graph. It must be mentioned at this point that the analysis of an assembly task using one criterion may lead to several sets of constraints which are alternatively valid (see example in Figure 16.2c).

Figure 16.3 shows a simple assembly task consisting of four different parts: a ground plate A, a pin B, a ring C and a cover D. If this task is analyzed according to the criteria stated above, the different constraint sets shown in Table 16.4 can be found.

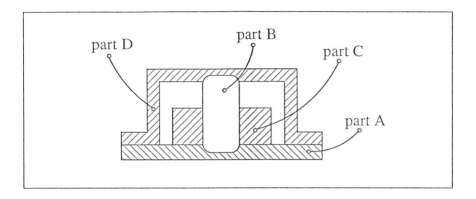

Fig. 16.3   A simple assembly task.

cause	resulting constraints set
rule 1	OR (AND($\pi(\vec{B},\vec{A})$, $\pi(\vec{C},\vec{A})$), AND($\pi(\vec{B},\vec{A})$, $\pi(\vec{C},\vec{D})$)), AND($\pi(\vec{B},\vec{D})$, $\pi(\vec{C},\vec{D})$)), AND($\pi(\vec{B},\vec{D})$, $\pi(\vec{C},\vec{A})$)))
rule 2	AND($\pi(\vec{A},\vec{B})$, $\pi(\vec{A},\vec{C})$, $\pi(\vec{A},\vec{D})$))
rule 3	AND(OR($\pi(\vec{B},\vec{C})$, $\pi(\vec{B},\vec{A})$)))

Table 16.4 The resulting constraint sets for a sample task.

The sets of constraints in Table 16.4 resulting from rule 1 and rule 2 are evident, while the set of rule 3 must be further explained: The assembly operation $\vec{B}$ may lead to problems if the parts A and C are already assembled because of a possible dislocation of part C. This fact can be expressed by the statement NOT($\pi(\vec{A},\vec{B})$) AND ($\pi(\vec{C},\vec{B})$) = TRUE. This is equivalent to $\pi(\vec{A},\vec{B})$ OR $\pi(\vec{B},\vec{C})$ assuming that NOT($\pi(\vec{X},\vec{Y})$) equivalent to $\pi(\vec{Y},\vec{X})$. This assumption is a simplification, since NOT ($\pi(\vec{X},\vec{Y})$) includes the possibility that the operations $\vec{X}$ and $\vec{Y}$ may be performed simultaneously. If this is to be considered, an attribute must be added to the constraints which represents the 'weight' of a constraint. With this factor, NOT $(\pi(\vec{X}, \vec{Y}))$ is equivalent to $\pi(\vec{Y}, \vec{X}, 0)$ OR $\pi(\vec{Y}, \vec{X}, 1)$ where the third value represents the weight. The constraint $\pi(\vec{Y}, \vec{X}, 0)$ means that there is no precedence constraint between $\vec{Y}$ and $\vec{X}$, but the constraint $\pi(\vec{X}, \vec{Y})$ results in a cycle and is therefore not allowed.

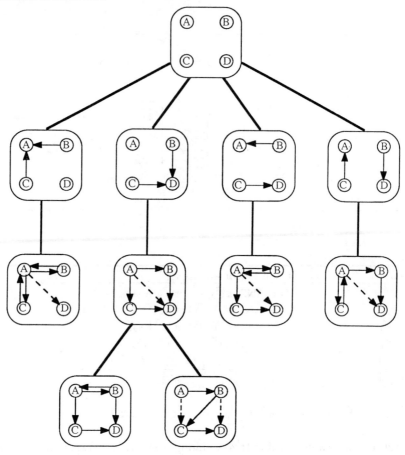

Fig. 16.5   Interference of constraint sets.

In order to integrate the constraint sets of Table 16.4 in one precedence graph, the operation set $\{\vec{A}, \vec{B}, \vec{C}, \vec{D}\}$ must be combined with one constraint set resulting from each rule. The result of this inference can be represented by a tree structure. Figure 16.5 shows the resulting graph for the sample task of Figure

16.3. The root (level 0) is represented by the operation set without any constraints. On the next level, the alternative constraint sets resulting from the penetration rule are included. The resulting four alternative graphs are then combined with the constraint set deduced by the stability rule. Since the precedence graphs must not have any cycles (because a sequence $\vec{X} \to \vec{Y}$ and $\vec{Y} \to \vec{X}$ cannot be executed at the same time) only one of the four graphs of level two is further combined with the constraint sets of the third rule (avoidance of side effects). On the last level, there are two resulting precedence graphs but only one graph is valid. It describes the resulting flow of assembly operations by the sequence $\vec{A} \to \vec{B} \to \vec{C} \to \vec{D}$ which is the only possible order of the assembly operations.

16.3.2 Automatic deduction of precedence constraints

The procedure of combining different constraint sets is quite simple. A more difficult problem is the automatic deduction of the constraint sets from a CAD model. This section describes in greater detail a module which analyzes a given assembly task using the penetration rule. On principle, there are two different approaches to solve this problem: The first approach consists of simulating the construction of the given assembly task. This approach leads to the problem of backtracking when an already assembled part is an obstacle for later assembly operations. This may lead to a time-consuming search process. The second approach (see also [9]) consists of disassembling the completed assembly. The precedence constraints valid for the disassembly are side-inverted with respect to the constraints valid for the assembly, and can therefore easily be transformed. In this way, backtracking can be avoided. The module presented here is based upon the second approach [10]. The basic version which was implemented requires the following conditions:

- All parts of the assembly have a constant shape during the assembly operation; i.e., parts like springs are not allowed;
- Every part of the assembly can be assembled in one straight movement. Combined movements consisting of translations or rotations are not considered;
- If a part can be disassembled in a given direction $\vec{r}$, it can also be assembled in the opposite direction $-\vec{r}$. Only one part of an assembly can be moved during one assembly operation.

Since these requirements are very restrictive, they may lead to constraints which are really not necessary. In this case the user has the possibility to interact with the system and to erase the undesired constraints. The principle of operation of the module can be subdivided into five different phases:

1. The description of all separate parts of the specified assembly is read, verified and then transformed into an internal object representation.
2. For every part of the assembly, a set of departure directions is determined.
3. This phase simulates the disassembly of the assembly task. Each workpiece is examined if it can be removed in one of its departure directions determined in phase 2. It is evident that the part which is (re)moved may not penetrate other parts of the assembly. During this process, the existing precedence constraints are computed.
4. All parts of the assembly which can be removed directly are now removed. It is important at this point to realize that not all precedence constraints are relevant for the precedence graph. What is vital is that each part must be removed at the earliest possible point of time. To facilitate this requirement, sets of parts which are removable at the same time are computed and then removed. This process is repeated until the whole assembly is disassembled.
5. The precedence constraints gained during the previous phases are now transformed into precedence constraints according to the assembly and represented in an adequate form.

16.3.3 Removal test

A precedence constraint between two parts results from the fact that one object at its goal position is an obstacle for the assembly of the other object. A test module was therefore designed which can decide if a part can reach its goal position without colliding with another part. If a part cannot be assembled, the obstacle which prevents this must be determined. The test module computes for every part the set of objects which are an obstacle during the removal. If an error occurs at this point, the error situation is displayed. In this way, the different precedence constraints for the assembly can be determined. As input, the module needs a description of the final state of the assembly task, the name of the part to be examined, and a selected departure direction for the part. The resulting output consists of the set of obstacles which prevent the removal of the examined part in direction $-\vec{r}$, and an error indicator. After the input phase, the test module computes for the part to be examined a set of suitable departure directions (phase 2). They depend on the geometry of all objects in the scene which are in contact with the object to be removed, called 'environment'. Removal directions for the object are as follows:

- the outward normal vectors of the contact faces of the environment;
- the sum of these vectors, and
- the two directions of each edge of all contact faces of the environment.

In the third phase, the module tests whether the different parts can actually be removed in the computed directions. This is done by the following procedure:

Consider an assembly consisting of several parts of the class 'polyhedra'. The part R is now examined to see if it can be removed in the direction $\vec{r}$. If there is a collision of R with an obstacle O, then it occurs between a face f_R of R and a face f_O of the obstacle. In order for this to occur, there must be a vector component of the outward normal of f_O in direction $-\vec{r}$ and a vector component of the outward normal of f_R in direction $\vec{r}$. In addition, f_O must lie in the path of f_R when moving in the direction $\vec{r}$. This problem can be reduced from a three-dimensional to a two-dimensional problem by means of projection. The face f_O, if lying 'above' f_R, is mapped by a parallel projection in direction $-\vec{r}$ into the plane defined by the face f_R. If O is an obstacle for R, then the projected face f^0_0 and the face f_R are overlapping.

Utilizing the test module, the disassembly of the given assembly task can now be simulated. If a part is removable in a given direction $\vec{r}$, then it has no predecessors according to the disassembly; that is, there are no other parts which must be removed before. The set of predecessors is empty. If the part is not removable in this direction, then the set of predecessors contains all the objects which are an obstacle for the part when moving in direction $\vec{r}$. If the test module is applied to every part of the assembly task and to every removal direction, then all information which is necessary to deduce the different precedence constraints according to the disassembly can be obtained. After this step the simulated disassembly is performed by removing all parts without a predecessor and updating the predecessor list. The computational cost can be minimized by reducing the list of removal directions by those directions which have a vector component in the opposite direction to the outward normal vector of any contact face of the environment. A removal of the object in those directions would result in a collision with a part of the environment. An example for the determination of the departure directions of an object is shown in Figure 16.6. Five directions were found for the removal of the bolt from the plate.

The calculated precedence constraints deal with relations according to the disassembly of the assembly task. They must now be converted into precedence constraints according to the assembly. The relationship is as follows:

139

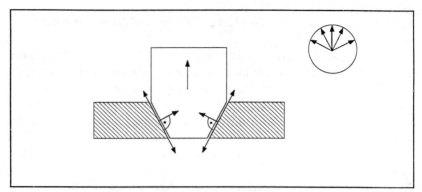

Fig. 16.6 Determination of suitable departure directions.

If part A is an obstacle to part B for its removal in direction $\vec{r}$, then A is also an obstacle for the assembly of B in direction $-\vec{r}$. A must be removed during disassembly before B, and during assembly, A cannot be assembled before B. B is therefore a predecessor of A and thus A is a successor of B by assembly of B in direction $-\vec{r}$. Since there are several possible assembly directions for each part, it makes sense to create a file for each part which gives all successors of this part for each assembly direction.

16.3.4   Estimation of computational costs

The computational cost of the test module can be determined from the number (n) of parts of the assembly task, the maximum number of faces per part (f_{max}) and the maximum number of edges per face (e_{max}). To determine all obstacles for a given object in a given direction, it must be compared with n-1 other objects. The comparison of two objects, in the worst case, requires f_{max}^2 comparisons between the faces. In most cases, it can easily be shown that the faces are not overlapping, e.g., by using a bounding box test. In all other cases, the projection and then the test for overlapping must be performed. This operation has the complexity of $O(e_{max}^2)$. From this an estimation of $O(nf_{max}^2 e_{max}^2)$ for one use of the test module is obtained. Since the test module must be performed for each removal direction and for each part, this amount must be multiplied with the maximum number r_{max} of removal directions per object. If only the direction of the vector sum of all outward normals of the touching faces is considered as removal direction, then r_{max} is a constant. If all directions of edges of touching faces are also considered, then r_{max}

evaluates to $f_{max} * e_{max}$ at the most. This leads to a maximum effort of $O(n^2 f_{max}^2 e_{max}^2)$ or $O(n2f_{max}^3 e_{max}^3)$, respectively.

16.3.5   System architecture

The object description is attained through the help of the CAD system ROMULUS of Shape Data, Ltd., and stored in a FEMGEN format [29]. The FEMGEN format allows the specification of arbitrary bodies. However, the implemented system requires that the objects are limited to the class of polyhedra. This means, in principle, no strong restriction, as bodies outside of this class can be approximated. For the specification of the final state of the assembly task, the ROSI system was used. During the first phase the analysis module converts the CAD into an internal model which is used to represent the geometrical and topological attributes of a three-dimensional body. A boundary representation was chosen which offers, through its hierarchical structure, good prerequisites for further processing of the object descriptions. Hereby, the body is specified by a set of bounding faces. Each face is defined by a set of bounding edges which are in turn defined by their endpoints. Topological and geometrical data of the objects which are not explicitly given though the CAD system must be calculated. In addition, a plausibility test of the CAD data is carried out.

16.3.6   The overall system

In the previous sections it was shown how precedence constraints can be deduced and how the constraint sets resulting from different analysis processes can be combined. The resulting precedence graph already gives an idea how the assembly sequence for the assembly should look like. This information is a necessary input for the subsequent detail planners, like layout planner, path planner, sensor planner, etc. The graph is therefore first transformed to a convenient representation called 'action-situation list'.

To demonstrate this, the simple precedence graph shown in Figure 16.5 is used as an example. When performing the operation $\vec{B}$, it is not defined whether the operations of the parallel branch $(\vec{C}, \vec{D}, \vec{E})$ have been already executed or not. Each of these objects may therefore be still at its start position (denoted by the index 0) or already at its goal position (denoted by the index 1). The situation for the operation $\vec{B}$ contains therefore the obstacle A_1 (because A is already at its goal position) and the obstacles C_0, C_1, D_0, D_1, E_0 and E1. The action-situation list for the whole graph is shown in Table 16.7.

action	objects in the scene
$\vec{A}$	B_0, C_0, D_0, E_0
$\vec{B}$	$A_1, C_0, C_1, D_0, D_1, E_0, E_1$
$\vec{C}$	A_1, B_0, B_1, D_0, E_0
$\vec{D}$	$A_1, B_0, B_1, C_1, E_0, E_1$
$\vec{E}$	$A_1, B_0, B_1, C_1, D_0, D_1$

Table 16.7 Action-situation list.

With the action-situation list as input, the detail planners have to plan the different actions for situations which are more difficult than in reality. They should now be able either to generate the parameters for the specified action and situation (e.g. motion parameters), or, alternatively, specify further precedence constraints. Figure 16.8 shows an example where a detail planner may find additional precedence constraints:

Two parts A and B should be picked up at their start position (left side) and then placed at their goal position (right side). The path planner has now to determine the motion parameters for the two operations $\vec{A}$ and $\vec{B}$ for a selected manipulator. Since there are no precedence constraints between the two operations, the situation for the operation $\vec{A}$ consists of the obstacles B_0 and B_1. The obstacle set for $\vec{B}$ is $\{A_0, A_1\}$. The path planner will not find a valid motion, since the gripper holding part A at its goal position will penetrate the part B at its goal position. From this it can infer that the motion is only valid if the obstacle B_1 is not present, i.e., if $\vec{A}$ is performed before $\vec{B}$. This can be expressed by the precedence constraint $p(\vec{A},\vec{B})$.

By this approach it is possible to maintain the degree of freedom of a graph representation until the execution phase. Each planning phase like task specification, task analysis, detail planning, and execution may find additional constraints according to the assembly sequence. These constraints are input for the central synthesis module which combines all constraints to find a valid order of the different assembly operations.

16.3.7   Conclusion

In this section it was shown how precedence constraints can be used to find a valid order of assembly operations. Such constraints may arise during every planning phase and therefore, they should be processed by a central synthesis module. Some classes of precedence constraints can be deduced automatically

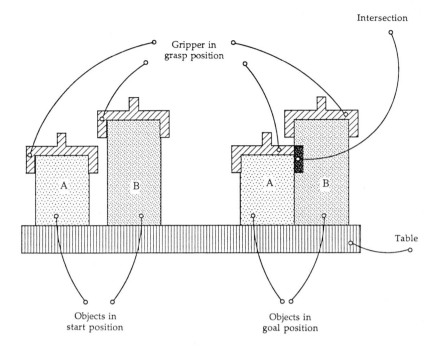

Fig. 16.8   A precedence constraint detected by the path planner.

from a CAD model of the assembly task. As an example, a module for detecting constraints according to the penetration rule was described. In addition, the overall structure of a planning system which operates on the basis of precedence constraints was presented. This system allows the integration of all planning phases and offers a unique interface to all subplanners.

16.4   Scheduling methods

16.4.1   Introduction

The problem of scheduling has been widely studied in the past, and many results were achieved, although these are, in most cases, valid only if a set of restricting assumptions are verified. This section presents the FIAR experience in ESPRIT Project 623 in tackling the scheduling problem of assigning assembly operations to the robots of a workcell. The approach selected is described after a brief introduction of the overall field of scheduling. Concluding the section are some

143

considerations on the methods that can be reasonably implemented for assembly problems given the feasibility constraints on execution time and computational power required by the scheduling activity. In general, the scheduling problem can be defined as follows: given

- n orders, each composed by a sequence of operations;
- m machines;
- the sequence of the machine applications to process each order;
- a cost function to evaluate each possible solution.

The problem is to describe the sequence of application of the machines on the various orders, assigning the starting processing time of each machine on the order to minimize cost function.

Various classifications of the scheduling problem can be defined; considering, for instance, the machine needed, single-machine or many-machine problems can be defined. For many-machine problems, flow shop problems are the ones in which each order passes through the machines in the same sequence. Other classifications can be defined.

To deal with scheduling problems, two main fields of methods are available. The first one is the application of Operational Research (O.R.) methods. The second is the application of Artificial Intelligence (A.I.) techniques.

O.R. methods can be divided into roughly two groups: methods based on complete enumeration in which the set of legal schedules is generated and tested against the objective function, and methods based on implicit enumeration in which the set of legal schedules is only implicitly generated and integer or dynamic programming techniques are used. The main disadvantages of applying O.R. methods are as follows:

1. Usually very restricting assumptions have to be verified which greatly limit the applicability of such methods to real cases.
2. The computational effort for real-world problems can be very high.

To partially overcome these problems, the application of A.I. techniques has been proposed in the past. One advantage is that A.I. techniques can deal with non-numerical constraints, as opposed to numerical ones, typical of O.R. methods. Moreover, A.I. methods can consider preferences and not only fixed constraints, and can vary (or relax) some of these constraints during operation. In many applications, the only approach that can be adopted is to use some A.I. techniques utilizing domain-specific knowledge, or heuristics, to reduce the search.

16.4.2 Characterization of the problem

The scheduling problem that FIAR studied within ESPRIT 623 has been, as previously introduced, the problem of assigning assembly operations to the robots of a workcell to carry out a given assembly task.

The module solving this task, also called Scheduler, is part of the expert system IMPRES (IMplicit PRogramming Expert Systems) which has been developed by FIAR in ESPRIT 623 to support the user in programming robots of a workcell for assembly tasks. For a complete description of the system, see Chapter 27. The general problem is the following: given

- the 'Task Description': the initial positions, before the assembly, and final postions, after the assembly, of all the workpieces, the type of contacts between objects in the assembled configuration and the geometrical description of all the workpieces and fixtures;
- the 'Cell Description': the model of the robots (kinematic parameters, definition of the robot language, geometrical description of the links) and the cell layout (robots' and fixtures' positions) determine the executable program for the robots of the cell which solves the assembly task.

Many problems have to be solved to achieve this goal. The main functionalities needed are as follows: Planning (to determine the Assembly Graph, that is, the order in which the parts have to be assembled), Scheduling (to assign assembly operations of the graph to the robots of the cell), Path Planning Generation (to generate the executable programs in the languages of the robot controllers). The rest of this section describes the approach taken to realize the second function. For the description of the others see Chapter 27.

A first solution adopted to reduce the complexity of the overall problem has been to split the overall scheduling problem in two. The first phase determines the Assembly Graph, that is, establishes some constraints between the operations. The result is a list of partially ordered operations, in which these are divided in groups (or steps), with precedence relations between them. An example of assembly graphs might be as follows: (STEP 1: assembly_1, assembly_4; STEP 2: assembly_5; STEP 3: assembly_3, assembly_ 2).

The assembly operations of each group can be executed in any order, and each group has to be fully executed before starting the following one. The scheduling problem has then been defined as the problem of assigning operations of each group to the robots, with the constraint of executing all operations of one group before starting to schedule the following one.

This approach has the advantage that, usually, the number A of operations on the same step is quite small (5-6 as a maximum); also, the number R of robots is usually small (2-3) and this is because IMPRES works at a cell- and not a plant level.

Complex assemblies (for instance, the assembly of a car) that require a plant with many cells must be previously divided by the plant engineer into cell-level subassemblies.

Because each operation of a step must be performed before each operation of the following step, each step can be considered independently from the others. This is actually a simplification which reduces the complexity of the problem, reducing also the optimality of the solutions found.

A tree is built by generating the possible assignments of operations to the robots; if a robot cannot perform an operation, because either the initial position or the final position of the workpiece is out of its working area, the assignment is discarded.

A search of a pseudo-optimum path is performed on this tree, using the A^* algorithm, estimating the time required for the transfer movements assuming linear trajectories (in joint-space) at maximum speed. This is the function g of the algorithm (remember that the function which guides the expansion of the states is $f = h + g$), while h is the time to actually execute the assembly operations obtained activating the Motion Planner to compute the collision-free trajectories for the robots. The function g is admissible, that is, gives a value always less than or equal to the real one, because in general the trajectories are not linear and not executable at maximum speed.

It can happen that the Motion Planner is not able to find a trajectory of this sequence, either because this trajectory does not exist or because it is too complex and a computational time limit is reached.

In this case the Scheduler backtracks, choosing a different path on the tree. This choice is done trying to save as many trajectories as possible, because the motion planning is quite a time-consuming job.

Of course, the method presented works only when the number of operations that can be performed in parallel is small: if this is not true, the system asks the operator to divide the operations into subsets in order to keep the size of the search space small.

As it is clear, there are a number of approximations of the scheduling solution found with respect to the optimal one.

This happens because finding optimal scheduling solutions requires a complex interaction between scheduler and motion planner with many backtrackings from the scheduling level and the motion planner level. Because both motion planning and optimal scheduling are time-consuming activities, an exact approach is not applicable or too high a computational time will be obtained.

16.4.3 Practical results

The experience gained with this system allows some considerations to be drawn on the trade-off between optimality of the generated robot program (minimum overall execution time) and computing time required for the scheduling activity.

The system has been tested on a set of examples. Among them, one (the 'Cranfield Benchmark' assembly) requires assembling 14 workpieces with 6 degrees of freedom PUMA260 robots using a fixture for the assembly and two pallets for the workpieces. Within this example the geometry is quite complex (about 2000 faces for describing workpieces and fixtures and 1000 faces for describing each robot). Conversely, the logical structure of the assembly is quite simple.

The CPU time required to obtain the program – on a DEC VAX Station – is less than 17 minutes, excluding time for graphic presentation and reading and writing of files. The complete execution, including graphical simulation of the various planning phases, requires about 75 minutes. As reported by the authors of the benchmark, the time required when the programmer is a human operator – with only one PUMA robot – is about 8 hours. These are clearly quite good results; similar performance can be expected for many assembly tasks.

More 'difficult' assemblies can require human intervention or a greater computing time; causes of difficulties are as follows:

- Assemblies with a very complex logical structure: this can require a lot of backtracking within the Scheduler. In this situation, the assembly task looks like a 3D puzzle; finding the correct order of the assemblies is required. This situation is not very common in an industrial assembly task, in which simplicity is preferred. When such situations occur, the best solution is to ask for human intervention.
- Assemblies requiring 'special' techniques - e.g., using ad hoc tools for assembling a workpiece, or using an intermediate fixture for building a subassembly. In this situation, the Scheduler by itself might not be able to find a solution. Of course it is not possible to have a system that knows all the techniques that have been - or that will be - invented; also in this situation the system needs the operator's help.
- Assemblies requiring planning of very complex trajectories: in this situation the motion planning time can be quite high. In the test described, about 30 trajectories are computed with a planning time from a few seconds to a few

minutes. Of course it is possible to find situations in which the motion planning time is greater. In these situations a time limit is applied to stop the motion planner; the operator can then decide to resume the search of the path or to help the system by defining intermediate robot positions. In some cases the cell layout might even be too cluttered and require too complex and long trajectories.

- Assemblies with a large number of workpieces that can be assembled in parallel: in this situation the scheduler can require a large execution time. Also in this situation the operator is involved in order to add other precedence constraints.

Concerning optimality of the solution found, the following considerations can be drawn: defining the optimal program as the program requiring the minimum execution time has little meaning because the time required to execute a program can be estimated with good precision only with complex dynamic simulations.

It is in fact usual to define the optimality in some approximated way (e.g., for a motion planner, the optimum trajectory is often assumed to be the minimum length trajectory in the joint space). Because of these approximations, it is not interesting to try to reach exactly the theoretical optimality; this usually requires very slow algorithms. A solution quite near to the optimal one is sufficient.

It is possible to draw some limited assumptions which, by reducing in an unpredictable but usually limited way the optimality, allow the activities to be divided into hierarchical levels. As an example, the scheduling activity is made up of three main hierarchical decision levels:

- Search in the scheduling space (space of the ordered sequences of assignments robot/operation);
- Search in the configuration space of the robot at reduced discretization (motion planner, high level);
- Search in the configuration space of the robot at high discretization (motion planner, low level).

This subdivision into levels allows the computational cost of the scheduling activity to be kept low. From a theoretical point of view, in order to reach optimality, a complete backtracking between the levels is required. But this implies the loss of the benefits in terms of computational cost of the hierarchical structure; if a complete backtracking is allowed, the hierarchical subdivision has no advantages anymore. Therefore, a complete backtracking cannot be allowed: a problem should be solved within its level as far as it is possible, possibly reducing the optimality of the solution.

148

This paradigm is typical, for example, of hierarchical organizations (such as a company) in which lower levels try to solve their own problems locally and report them to the upper levels only when it is necessary. This can reduce the effectiveness of the organization (because local solutions can be less optimal than global solutions) but this is the only way to have a manageable structure.

The conclusion is therefore that A.I. methods applied to scheduling can reach in a limited time, a solution which is not the optimal one, but 'close' to it, by dividing the problem into hierarchical levels for cases in which O.R. algorithms are not applicable.

16.5   Hierarchical collision-free path planning

16.5.1   Introduction and state of the art

A path planner is a system able to find a collision-free path beetween two given positions in an environment containing obstacles. Usually, the path is defined by a sequence of intermediate positions (corner points).

A further step – trajectory planning – is required for defining the motion between the corner points. In particular, the following items have to be decided:

- type of motion (usually linear joint interpolated motion);
- speed;
- acceleration;
- smoothing at corner points.

At present, the most commonly used approaches are based on the 'Configuration Space' (C-space) concept [11] [12].

The C-space of a manipulator with n degrees of freedom is a subset of an n-dimensional space. In other words, each degree of freedom of the robot gives birth to a dimension of the C-space.

Each configuration of the manipulator, defined by the joint coordinates $(q_1,...,q_n)$, is represented by the points in the C-space having $(q_1,...,q_n)$ as coordinates. The C-space is bound by the joint limits values; therefore, the C-space is an n-dimensional parallelepiped. A point of the C-space is forbidden, with respect to a set of obstacles, if a link or the payload of the manipulator intersects an obstacle. The Forbidden Space is defined as the set of all the forbidden points. The Free Space is defined as the intersection between the C-space and the Forbidden Space.

The path planning problem can therefore be fomulated as the problem to find, when it exists, a continuous line contained in the Free Space connecting two given C-space points.

The classical approaches are based on a two-step procedure:

1) Compute a representation of the Free Space.
2) Find the path.

The Free Space, which is in general not convex, can be represented as a set of overlapping or disjoint convex Free Sets plus a connectivity graph. The connectivity graph nodes are the Free Sets; a (not oriented) link connects two adjacent Free Sets. The path-finding activity requires the following:

2.1 Finding the node of the connectivity graph to which the start position belongs; this is called the start node (if the start position does not belong to any node, the start position is not collision-free and therefore no collision-free path exists);

2.2 Finding the node of the connectivity graph to which the goal position belongs; this is called the goal node (if the goal position does not belong to any node, the goal position is not collision-free and therefore no collision-free path exists);

2.3 Finding a path on the connectivity graph betwee the start and the goal node (a connected sequence of links). If no such path exists, no collision-free path exists;

2.4 For each link on the path, finding a C-space point that allows the transition between the two adjacent Free Sets. This is a point on the intersection (if the Sets are overlapping) or on the common boundary (if the Free Sets are disjoint) of the two Free Sets.

Because the Free Sets are convex, the linear (in the C-space) path between any pair of points belonging to the same Free Set is always collision-free. Therefore, the set of C-space points defines a collision-free path.

Step 2.4 requires some optimization on the choice of the point on the common boundary in order to obtain a globally optimized path. Moreover, step 2.3 requires an optimization of the choice of the path.

As an example, Lozano-Perez performs step 2.3 using the A* search algorithm. Step 2.4 is executed during the A* search: from the current position, the point on the boundary closest to the current one is chosen, and the distance (in the C-space) between these two points is used as cost function. This requires, of course, to have a representation of the common boundary associated with each link.

The difference between the various approaches is mainly due to the representation of the Free Space sets: to have a compact representation, the spatial coherence of these sets has to be used. In order to do this, Lozano-Perez uses regions and Faverjon uses octrees (see [11] [12]).

The approaches based on C-space are quite fast as far as the path-finding acitivity is concerned: the dimension of the connectivity graphs is quite low and the search is fast. Moreover, these approaches work well for very cluttered environments as well.

Unfortunately, building the Free Space is computationally a very expensive activity, growing in an exponential manner with the number of degrees of freedom. When the environment is static – that is, the position of the objects does not change – the Free Space computation can be done only once. But when the positions of the obstacles change, or when the geometry of the manipulator changes (because a payload is grasped or released, as a typical example), the Free Space has to be recomputed. This is due to the following:

- The transformation of obstacles from the Cartesian Space representation to the C-Space representation depends strongly on the manipulator geometry;
- The C-Space representation of an obstacle after a translation and/or rotation of the obstacle changes in an unforeseeable manner and it is therefore not possible to obtain the new C-Space representation from the old one.

Moreover, the computation of the Free Sets depends on the manipulator kinematics and geometry: an ad-hoc algorithm for each manipulator has to be defined.

16.5.2   The basic algorithm

Given the constraints typical of assembly problems, and specifically that

- in an assembly task, the environment is not static (each assembly operation changes it);
- in an assembly task, the geometry of the manipulator changes from trajectory to trajectory (the robot grasps and releases the objects to be assembled);
- an off-line system has to be easily reconfigurable with respect to the manipulator geometry and kinematics;
- in most of the situations where the environment is not very cluttered, a different approach has been chosen.

151

The C-Space can be discretized as a n-dimensional grid of equal cells, each of them representing a discretized position of the n degree of freedom manipulator.

Each cell is adjacent to $(3^{**}n-1)$ cells: this means that the discretized C-space can be considered as a graph.

The start position and the goal position belong to two cells: the starting cell and the goal cell.

An A* algorithm is used in order to find the optimum path (optimum means the shortest C-space path).

The cost function is the length of the already defined path.

The estimate function is the distance (in C-Space sense) between the current cell and the goal cell.

A cell is checked for admissibility finding the Cartesian positions of the links and payload (by means of a forward kinematic transformation) and checking (in the cartesian space) intersections between the manipulator elements (links and payload) and the obstacle. If an intersection is found, the cell is not admissible: it is forbidden.

It is straightforward to observe that, when all the cells are admissible, i.e. no obstacles exist, this algorithm finds the linear path in the C-Space.

It is not necessary to have an explicit representation of each cell: only the cells examined during the search are generated.

It can be noted that this method is similar to the C-Space based ones, with two main differences:

Not all the Free Space is generated, only the part required during the search: when a cell is checked for admissibility, it is found if the cell belongs to the Free Space or to the Forbidden Space.

The Forbidden Space representation is simply the list of the forbidden cells, without coherence exploitation.

The complexity of this problem depends on the following:

- the complexity of the search graph: this depends on the chosen resolution and on n (where n is the number of degrees of freedom):
- the number of cells of the graph is $N^{**}n$, where N is the number of discretized ranges in which each coordinate axis is subdivided.

Choosing a little value for N, the complexity of the algorithm can be kept quite low. Unfortunately, this is not possible. In fact, because the path is checked only at discretized positions (the cells), the path between two cells cannot be collision-free: if the discretization is low, objects causing collisions when the path is executed might be missed when the latter is planned.

It has been demonstrated that this problem can be solved enlarging the obstacles by the cell dimension. This poses a limit on the dimensions of the cells: if the dimension of the cells is too large, the obstacle enlargement can prevent the system from finding an actually feasible path.

Therefore, N shall be quite large and consequently the complexity of the algorithm (both memory for storing examined cells and time for checking the cells for collision) is high. A method to reduce this problem is described in the following.

The main advantage of this algorithm is its reconfigurability: relying only on forward kinematics, and Cartesian space collision checks, the kinematic model can be easily defined by means of the Denavitt-Hartemberg parameters, the robot (and obstacles) geometry can be defined using a CAD system.

16.5.3 The two levels of the algorithm

In order to reduce the complexity of the algorithm, a two-level system has been developed.

Both of the two levels rely on the A* algorithm: the first one finds a path on a n-dimensional grid with low resolution (cell dimension quite large); the second level finds paths between the corner points found by the first level.

The admissibility on the first level is checked by the second level. This allows the A* algorithm to be applied on a graph of limited dimensions: for example, if the required resolution imposes the subdivision of the space into $(100)^{**}n$ cells, the first level can perform a search on a graph composed by $(10)^{**}n$ large cells; the second level performs the search between the center of these large cells, subdividing this region into $(10)^{**}n$ little cells.

Because the complexity of the algorithm grows with $N^{**}n$, this method allows the computational time required to be kept reasonably low. This important advantage is obtained at the price of planning sub-optimal paths. The experience has shown that, usually, the loss of optimality is quite little, but this cannot - of course - be always true.

Another method used to keep complexity low is to use, instead of a pure A*, a weighed A*; the weighted A* uses a weight $0=<W<=1$. When $W=1$, the weighted A* coincides with A* and finds optimal paths in the graph; when $W=0$ the weighted A* degenerates in the so-called heuristic search that finds a non-optimal path but is sensibly faster.

Intermediate values of W allow making a trade-off between optimality of the path and required computational time. This simple trade-off can be done by the user, if necessary.

Experience has shown that, usually, choosing W equal to about 0.8 allows finding paths quite near to the optimal one with a sensible improvement of the computational time performance.

Concerning the loss of optimality, it is important to note that the optimality, as reached by a one-level system with W=1, has only a geometric sense: the path is the shortest one in the C-space. This does not mean that the path is optimal with respect to the execution time: this second kind of optimality can be reached only by a system that takes into account the dynamic behaviour of the manipulator as well.

In general, the optimal path, in the geometric sense, has a lot of corner points that have to be smoothed in order to be dynamically feasible; this smoothing requires deceleration and acceleration, and therefore loss of execution time. Therefore, some limited loss of geometric optimality has a quite reduced impact on the dynamic optimality.

16.5.4 Checking for intersections

The method used to perform collision checks in the Cartesian space is based on a geometric model in which the obstacles, the manipulator links and the payload (called elements) are represented as sets of convex polyhedra. A pre-processing program, able to convert a standard CAD output format (FEMGEN) into FIAR internal representation, splitting non-convex polyhedra into convex ones, has been developed.

An algorithm able to find quite rapidly if two convex polyhedra have a non-empty intersection has been defined.

In order to avoid calling this algorithm for a lot of pairs of convex polyhedra at different times, three levels of preliminary checks are used:

1) The sphere circumscribing two elements (for example, a manipulator link and an obstacle) are checked for intersection. However, this is very simple, and a fast check cannot exclude intersection if the second step is performed.
2) The bounding boxes (i.e., circumscribing parallelepipeds) of the two elements are checked for intersection. This check is performed by the same algorithm that finds if there is intersection between two convex polyhedra. Only if this rapid check cannot exclude intersection is the third step performed.
3) For each part of convex polyhedra from the two sets of polyhedra composing the two elements, the spheres circumscribing the convex polyhedra are checked for intersection. Only if this check cannot exclude intersection is the intersection check between the two convex polyhedra done.

The intersection between two convex polyhedra is checked using a simplex-based algorithm.

A convex polyhedron can be represented as the intersection of the halfspace defined by the faces of the polyhedron; the intersection of two polyhedra is therefore defined by the intersection of the halfspaces of both the polyhedra. This defines, in the Cartesian space, a region R limited by linear constraints.

An intersection between the two polyhedra exists if and only if, R is not empty; that is, if there exists an admissible point satisfying all the constraints. This is the classical initialization problem of a linear programming problem and can be solved by one of the well-known simplex-based methods.

16.6 Grasp planning for a three-fingered hand

16.6.1 Introduction

In an Implicit Programming System, after having decomposed the task into a sequence of actions, an action execution planning phase must start, which must be performed by some specialized modules such as the grasp planner, the fine motion planner, and the gross motion planner. In the following, the design of a grasp module to be included as a subcomponent of the 'Planning System' is described. Such a module deals with the problem of finding the grasp points and the contact forces needed for a stable three-fingered hand grasp of a given object.

Starting from the definition of the grasp problem, an abstract model for grasping is defined and a possible algorithm which implements it is shown. This algorithm is based on a rule system used to describe both the physical and the heuristic knowledge suitable for the grasp planner.

The overall study of the grasping problem is an open and complex problem which requires considering various situations. The identification of some behavioural constraints can be useful for the planning algorithm.

The best approach starts by considering human comprehension. As a first step, two main phases can be distinguished: the preshape phase and the grasp execution phase.

During the preshape phase, the location and orientation of the target object is identified, the appropriate degrees of hand freedom are selected, the contact surfaces are determined, and the approach trajectory and the opening of the hand to a dimension larger than the one of the target object are selected [13]. The knowledge about the task, the object to be grasped, and the state transitions have to be structured so that the design of a proper task description knowledge-base and of an object representation paradigm simplifies the generation of the appropriate set of grasp preshapes.

Many classifications have been proposed for grasping. For example, three main grasps are identified in Lyons [14]: encompass, lateral, and precision, capturing the interconnection between shape and functionality. A more various taxonomy of manufacturing grasps is considered in Cutkosky [15], together with an expert system for the selection of a preshape among them. Only precision grasps are considered, i.e., grasps that allow a maximum amount of possible manipulation by translational and rotational movements of the three fingertips.

When the grasp is executed, the hand comes into contact with the object and the forces which are necessarily exerted. Therefore, a correct set of grasp points has to be chosen, a safe hand configuration among surrounding obstacles must be computed, and it must be ensured that the forces and torques applied by the fingertips can compensate the external forces and torques. In addition, other constraints must be satisfied such as, for example, realizing the suitable contact between the object and the hand fingers.

The analysis of a legal grasp can be usefully limited to verifying the fulfillment of the basic requirements that the grasp must be reachable and must be stable [16] [4]. The grasping problems can therefore be clarified by considering the various constraints that affect reachability and stability.

It is useful to consider the following classifications of reachability: local, static, and dynamic.

Local reachability is determined by the constraints given by the relationship between the shape and dimension of the hand and of the objects. The static reachability is determined by the constraints given by the surrounding objects. The dynamic reachability is determined by the constraints given by the grasped object while they move among obstacles.

Information about surfaces is basic for reachability: definition of objects in terms of solid geometry can give useful information for global planning reasoning, but only the surface details can help in detecting the fingertip contact points. In Mirolo [17], an algorithm has been developed to detect collisions among convex polyhedra to use it in a local planning approach.

In the analysis of stability three levels can be considered: equilibrium, force-closure, and proper stability.

In an equilibrium grasp the sum of all contact and external forces is zero.

Through a force-closure grasp, arbitrary forces and torques can be exerted through the grasp points. Equivalently, the object cannot break its contact with the finger tips unless some non-null external work is exerted [17]. Of course force-closure is not a necessary condition for equilibrium grasp. It is very useful to consider that some grasps, having the sum of all contact forces equal to zero, are also force closure grasps [19].

The characteristic of a stable grasp is that it returns to its original position when the object is displaced by small external disturbances. Stable grasp can have an active stiffness control of the hand and the grasped object. In [20], it has been shown that all 3D force-closure grasps can be made stable.

In the algorithm, only force-closure grasps are considered [21]. The geometric modeler WM (World Modeler) was used as a modeling system to develop the grasp planner, because it provides the required information [22]. WM was developed at the Institute LADSEB of CNR and has been extensively used for a few years to develop robotics cell simulators. The information required to build 3D models in WM can be easily obtained from the other modelers developed within the ESPRIT Project by using the 'FEM GEM' interface protocol. In the WM, a model of the Salisbury hand was built (see Fig. 16.9) which was used for the development of the grasp planner.

16.6.2 An abstract model for grasping

The contacts between hand and objects are limited to the fingertips. Contacts between fingertips and objects can be modeled preferably as frictionless point contacts, hard-finger contacts, or soft-finger contacts. Only hard-finger contacts are considered, i.e. point contacts with friction. In hard finger contacts, the finger can exert any force pointing into the friction cone at the point of contact.

A possible formal represention for the geometry of contacts is obtained by using the screw representation of forces and velocities. For each contact it can associated both a set of lines on which generalized forces (wrenches) can be exerted, and another set of lines on which generalized displacements (twists) can take place [23]. By the wrench and twist representation, it is easy to understand the geometry of constraints and to introduce the data for the contact matrix which solves the force closure problem. In the following, as long as possible, an informal description of problems was used.

In the following, it is considered how reachability and stability can be characterized to obtain the set of tests that are to be introduced in the grasp reasoning system under the form of production rules.

(a) Reachability

(i) Object shape and approach directions. The object's form influences the hand preshape. Since the constraints are determined by the possible collisions among object surfaces while objects move, it is convenient to keep complete information about object boundary elements. Object volumes on the contrary give useful information for evaluating world model consistency. Distinction

157

should be made between what has not yet been decided during the planning phase and what can never be known (due to parts tolerances).

The choice of a particular representation scheme, like Wire Frames, Generalized Cylinders, Octrees, or others, influences the design of the planning algorithms.

The existence of invariants with respect to object position and orientation, a quick accessibility to object surface information, the possibility of ordering the solid representation elements in such a way to allow for a fast search among them (thus yielding an anisotropic representation), and the decomposability of a complex object into simple solids are all significant in determining the complexity of the algorithms.

In the C-Space approach, for example, objects are described by specifying which volumes of a predetermined space subdivision are filled or empty so that invariants are lost. Changing object position or orientation therefore affects all representation items. To avoid the above problem, it is necessary to keep the information on object structure separate from the information on object location.

Most spatial reasoning algorithms work efficiently if the represented geometric structures show some preferred direction, in contrast with the fact that in the abstract geometric space objects may be represented without giving prominence to some particular direction (unless some physical external constraint need to be added, like the gravity force). Therefore the information about the solid surface local orientation has been introduced explicitly in the geometric modeler WM.

Much attention must be given to reasoning about local geometry properties, since global planning algorithms about robot motion are often the result of too many simplifications in the abstract assumptions about the real world, or they are computationally very expensive.

Building the Configuration Space, for example, may result in not preserving the original detailed topological information about space obstacles. Polyhedra obstacles may give origin to nonplanar surfaces.

At the beginning, global techniques can be useful to decompose the problem and reduce the dimensions of the abstraction space. Global optimization algorithms allow fast tests for collision along pre-specified paths and arm gross motion. But a correct interpretation for object relationships must be found directly in 3D geometric world model by local planning techniques. Robot local planners must make good use of the information given by the internal data structure of the supporting geometric modeling system. In this way it is unnecessary to recompute the search space each time a modification occurs, as in the C-space approach.

(ii) Collision detection between convex polyhedra. The key element of the approach presented here is the introduction of the surface information. This constitutes the main constraints generated by the physical environment during the model evolution.

In Mirolo [24], an algorithm for fast detection of collision between convex polyhedra was presented. The algorithm, detecting the collision of a 3D rigid polyhedral body moving towards a polyhedral obstacle, can be formally expressed in the following way:

Given a consistent configuration of two convex polyhedral rigid bodies, S1 and S2, and an oriented direction d, parallel to the generating section of S1, compute how much must S2 be translated along d in order to bring S2 into contact with S1.

The algorithm allows the implementation of an operator with a meaningful interpretation relative to the geometric model. The effect of its application can be the possibility of moving one polyhedron along a prefixed direction up to a contact configuration, with a second polyhedron, as clarified by Fig. 16.10.

Only the obstacle surface elements that are strictly necessary need to be computed. Polygonal surfaces are represented as anti-clockwise sequences of Cartesian points. Successive configurations of generating polygonal sections are connected maintaining the orientation of the spine. Parallel sections are the source of information for faces and drums, and face normals are pointed by their corresponding vertices.

Polygon sides and polygonal faces therefore may be easily ordered by considering their normals. Since convex polyhedral solids show regular surfaces, considering successive segments of a sweep rule given as a broken line makes it possible to order a sequence of different drums generated by the section.

If a possible hand preshape has been pre-determined, and if a face triplet which gives a force-closure grasp for a large sub-area in each face has been found, the fingertips can be guided along their approach directions to find the proper force application points.

(b) Stability

The problem of verifying grasp stability was split into two steps: first the selection of three points of contact for which a force-closure grasp exists, second the computation of the grasp forces necessary to contrast the external forces. The approach of Nguyen [18] was used for the first step, and the result given in Kerr [25] for the second one. In the system, the hand approach direction is constrained to be orthogonal to the plane belonging to the three fingertips. A grasp plane is defined by the triplet of object points which are candidate to be contact points.

159

(i) Force closure problem. Starting point was the definition of a non-marginal equilibrium, with all contact forces strictly directed within their respective friction cones, as the case in which a strictly positive sum of the contact wrenches is zero.

Therefore given three grasp points, each one realizing a hard-finger contact and satisfying the previous restricted equilibrium condition, in Nguyen [19], it is proved that the resulting grasp is force-closure. From a formal point of view, verifying force-closure corresponds to verifying if a system on n linear inequalities, where matrix rows are the contact wrenches, has no solutions.

In Nguyen [19], it is also suggested to split the force-closure problem into two independent subproblems, force-direction and torque closure. Because of the independence between the subproblems, it was found to be particularly suitable to implement the force-direction closure tests in a rule-based reasoning system by using the surface information offered by our geometric modeler WM. In WM, 3D objects are represented as generalized cylinders, restricted to convex polyhedra with each section parallel to the generating section and normal to the spine. The information about local orientation of the solid surface is kept explicit by pointing out face normals through their corresponding vertices.

Therefore, first a procedure was built which tests the direction-closure part of force-closure, i.e. it verifies if there exists a vector triplet, internal to the friction cones, which spans all directions in 3D. Because faces in WM objects are planar, the direction-closure of forces is independent from their application points.

A second procedure allows the testing of the torque-closure condition which is true if the torque sum is equal to zero. For convex polyhedra this happens if and only if the three force directions have an intersection point; moreover, the forces are in equilibrium if they are coplanar and their application points do not belong to the same line. This result leads to a further constraint which imposes that the forces have to lie on the same plane in order to satisfy the force-closure condition.

(ii) Force setting. Once the three contact points are determined, the system automatically computes the contact forces necessary to balance external forces by using the model given in Kerr [25].

Because the Coulomb friction model gives non linear friction constraints, friction cones are approximated to convex polyhedra. Friction constraints and joint torque limits are therefore modeled by a set of linear equations. Force and torque balance equations relate the external environmental forces and torques applied to the object, to the components of the forces and torques applied to the object by all fingertips.

From these equations and constraints, a linear programming system solvable by the Symplex Method is obtained. As a function to maximize, the one given in Kerr [25] was chosen, which yields the maximum value for the minimal distance d; d is the distance (in the solution space of the linear programming problem) between each candidate solution (the magnitude of the internal grasp forces) and the planes given by friction and torque limits.

16.6.3   The grasp planning algorithm

Starting from the previous considerations, a procedure was constructed to find out a set of suitable grasp configurations by combining reachability and stability tests over the set of all admissible contact face triplets.

To each candidate triplet, a state is associated which describes its condition, as clarified by Figure 16.11, plus a set of features that gives additional information regarding various properties of the grasp determined by that triplet. Special attributes are used to characterize how good the grasp is. They take care of particular characteristics that give useful advice about which grasp to prefer, because, as it is known, there is generally not a unique stable and reachable grasp on an object. Therefore, the choice can be guided by such criteria as a smaller dependence from friction, and a greater distance from the obstacles.

Considering the triplet of faces allows to reason at a symbolic level about the set of all the possible grasps on an object [26]. Each triplet represents the equivalence class of all the grasps with contact points belonging to those three faces.

Some heuristics based on some special features of 3D object properties are useful to find a stable grasp in a fast and easy way. In the case of two overlapping parallel faces, for example, i. e. when the projection of one face on another has a non-null intersection, it can immediately be asserted that for all grasps which have two points on one of these faces and the third opportunely placed on the other, the force-closure property holds. In the case in which the center of gravity of the object has projections inside each triplet face, the normals to the faces of the triplet belong to the same plane and the force-direction-closure property holds, then the force-closure testing is straight forward.

This algorithm is suitable to be represented through a set of production rules in an expert system, communicating with conventional applications for numerical computation. The rules can codify the searching strategy for the grasp; the values of features and attributes guide the choice of the appropriate test for each triplet of faces. A rule is fired when a triplet assumes a particular state related to this rule; in fact, using rules gives the system a certain degree of flexibility because the order of test execution is not fixed in time.

161

The set of rules can be divided into several groups, depending upon the aspect they are related to: stability, reachability, force computation, functionality, and so on [27].

An object-oriented approach using the concepts of class and instance is suitable to represent the physical and geometric model of objects and three-fingered hand used by the algorithm for computing the grasp points.

The system architecture is based mainly on three modules: the world modeler, the grasp planner and the control module for the three-fingered hand, which communicate with each other on a message-passing basis.

The world modeler WM is used to define and maintain the geometry and physical characteristics of the objects to be manipulated, and the 3-fingered hand. It also acts as a geometrical data-base for the other modules, answering the queries about properties of the objects and relations among geometrical features, such as faces, edges and points.

The grasp planner represents the knowledge used in grasping an object. Based on a rule system, it computes the grasping points on the object's surface and the position of the hand. Moreover it verifies some constraints regarding the reachability and stability problems.

Finally, the control module for the three-fingered hand implements the kinematics equations used in controlling the position of the hand. It is based on the Salisbury's hand and communicates both with the world modeler for the graphic visualization and with the grasp planner from which it receives the grasping points and the hand's posture. This last information is given through a specialized language which allows the description of the hand's operation by means of a linguistic structure [28].

16.7 Conclusions

In this chapter, the overall design of a module for grasp planning has been presented that has been thought to be included as a module in the general Implicit Programming System developed by the ESPRIT Project No. 623. The module itself is complete and can be interfaced to the other module of the project by exchanging the geometric data through some CAD protocol, like FEM GEM, or others. The stability problems have been solved for the simplest case given by a model of hard-finger contacts in force-closure equilibrium. The reachability problems have been abstracted in a world built by 3D convex polyhedra obtained through the geometric modeler WM. For this representation, a very fast algorithm for collision detection has been presented. This algorithm enables solving the local geometric reasoning part of the reachability problem. The same global geometric reasoning techniques developed by the project for the other modules such as the path planning algorithm developed by FIAR can help solve the other part of the reachability problem. The software is actually running: the collision algorithm has been implemented on a Sun Workstation at the University of Udine, in the case in which the two approaching polyhedra have their generating sections both parallel; a set of stability tests has been written as production rules of the shell NEXPERT at the Institute LADSEB. Our system will be further improved and enriched both theoretically and experimentally in the ESPRIT-2 Project CIM-PLATO that will follow the ESPRIT-1 Project No. 623.

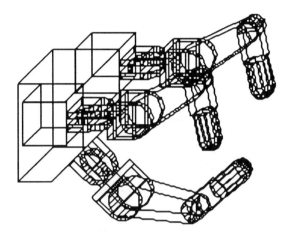

Fig. 16.9   The model of the salisbury hand.

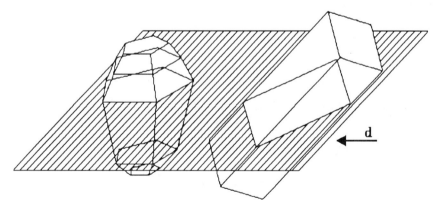

Fig. 16.10 Two polyhedra approaching along the direction d.

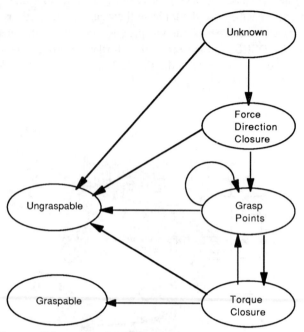

Fig. 16.11 The possible states of candidate triplets.

References

[1]   T. Lozano-Perez and P.H. Winston, *LAMA: A language for automatic mechanical assembly*, Proc. 5th IJCAI, Cambridge, Mass, Aug 1977.

[2]   L.I. Liebermann and M.A. Wesley, *AUTOPASS: An automated programming system for computer-controlled mechanical assembly*, IBM Journal of Research and Development, Vol. 2, No. 4.

[3]   T. Lozano-Perez and R.A. Brooks, *An approach to automatic robot programming* in *Solid Modelling Applications* (eds. J.W. Boyse and M.S. Picket), Plenum Press, New York.

[4]   T. Lozano-Perez, J.L. Jones, E. Mazer, P.A. O'Donnel,W.E.L. Grimson, *Handey: A robot system that recognizes, plans, and manipulates*, Proc. of the IEEE Int. Conf. of Robotics and Automation. Raleigh, N.C. 1987, pp. 843-849.

[5]   R.P. Paul, *Robot Manipulators: Mathematics, Programming and Control*, The MIT Press, Cambridge, Massachusetts and London, U.K., 1981.

[6]   S. Bonner and K.G. Shin, *A Comparative Study of Robot Languages*, Computer, IEEE,1982.

[7]   R.J. Popplestone, *Specifying Manipulations in Terms of Spatial Relationships*, DAI Research Paper No 117, International Seminar on Programming Methods and Languages for Industrial Robots, 27-29 June, IRIA Roquencourt, France, 1979.

[8]   W.J. Clesle, *Interpretation of spatial relationships among 3D objects* (in German), 'Studienarbeit', Institute for Real-Time Computer Control and Robotics, University of Karlsruhe, Germany, 1987.

[9]   T. Sata, F. Kimura, H. Hiraoka, H. Suzuki and T. Fujita, *Comprehensive modelling of a machine assembly for off-line programming of industrial robots* in *Off-line Programming of Industrial Robots*, (A. Storr and J.F. McWaters, Eds.), Elsevier Science Publishers B.V. (North Holland), IFIP, 1987.

[10]   J. Hornberger, *Ein Verfahren zur automatischen Ableitung von Montage-Vorrangbeziehungen aus einem CAD Modell*, Master thesis, University of Karlsruhe, Institute for Real-time Computer Control Systems and Robotics, Germany, 1987.

[11]   T. Lozano-Perez, *A simple motion planning algorithn for general robot manipulators*, AAAI-86, Philadelphia, 1986.

[12]   B. Faverjon, *Obstacle avoidance using an octree in the configuration space of a manipulator*, IEEE Int. Conf. on Rob. and Auto., Atlanta, 1984.

165

[13]   R. Tomovic, G.A. Bekey, W.J. Karpus, *A Strategy for grasp synthesis with multi-fingered robot hands*, Proc. of IEEE Int. Conf. on Robotics and Automations, Raleigh 1987, pp. 83-89.

[14]   D.M. Lyons, *A simple set of grasp for a dexterous hand*, Proc. of IEEE Int. Conf. on Robotics and Automation, St. Louis, April 1985, pp. 588-593.

[15]   M.R. Cutkoksky, *On grasp choice, grasp models, and the design of hands for manufacturing tasks*, IEEE Transactions on Robotics and Automation, Vol. 5, No. 3, June 1989, pp. 269-279.

[16]   T. Lozano-Perez, R.A. Brooks, *An approach to automatic robot programming*, A.I. Memo No. 842, MIT-AI Lab, April 1985, pp. 1-35.

[17]   C. Mirolo, E. Pagello, *Motion planning algorithms for mechanical assemblies*, Reprints of the IFAC Symposium on Robot Control-SYROCO 88, Karlsruhe, October 1988, 42.1-42.6.

[18]   V.D. Nyugen, *Constructing force-closure grasps*, The International Journal of Robotics Research. **Vol. 7**, No.3, 1988, pp. 3-16.

[19]   V.D. Nyugen, *Constructing force-closure grasps in 3D*, Proc of IEEE Conf. on Robotics and Automation, Raleigh 1987 pp. 240-245.

[20]   V.D. Nyugen, *Constructing stable graphs in 3D*, Proc. of IEEE Int. Conf. on Robotics and Automation, Raleigh 1987, pp. 234-239.

[21]   P. Bison, E. Pagello, L. Stocchiero, *A grasping module for task level robot planning*, Working Paper, ESPRIT Project No. 623, IP-CNR-1.89/1, Padova, January 1989.

[22]   C. Mirolo, E. Pagello, *Characterization of solid modelling properties for robot motion planning*, Working Paper ESPRIT Project No. 623, IP-CNR- 12.87/1, Padova, December 1987.

[23]   J.K. Salisbury, *Kinematics and force analysis of articulated hands*, in M.T. Mason and J.K. Salisbury, *Robotic Hands and the Mechanics of Manipulation*, The MIT Press, Cambridge, 1985, pp. 1-94.

[24]   C. Mirolo, E. Pagello, *A solid modelling system for robot action planning*, IEEE Computer Graphics and Applications, January 1989, pp. 55-69.

[25]   J. Kerr, B. Roth, *Analysis of multi-fingered hands*, The International Journal of Robotics Research, Vol.4, No. 4, 1986, pp. 3-17.

[26]   J.P. Troccaz, *Geometric reasoning for grasping: a computational point of view*, NAOT ASI Series, Vol. F30. CAD-Based Programming for Sensory Robots (ed. Bahram Ravani), Springer Verlag, Berlin, 1988, pp. 397-423.

[27]   P. Bison, C. Ferrari, E. Pagello, L. Stocchiero, *Using sensor data and a priori knowledge for planning and monitoring multi-fingered grasping*, International Workshop on Sensorial Integration for Industrial Robots. Zaragoza, November 1989.

[28]   P. Bison, E. Pagello, G. Trainito, *VML: An intermediate language for robot programming*, Robotics and Computer Integrated Manufacturing, Vol.5, No. 1, January 1989, pp. 11-19.

[29]   E.R. Argyle, *ROMULUS to FEMGEN Interface, The current System*, Revision 2, Shape Data Ltd., April 1984.

Chapter 17

Exception handling

G.R. Meijer, University of Amsterdam, The Netherlands
V.Caglioti, M. Somalvico, Politecnico di Milano, Italy
U. Negretto, University of Karlsruhe, Germany

17.1 Evolution of robot systems

This chapter deals with the problem of handling unforeseen situations during the execution of a robot program. Traditionally, these situations would lead to a complete standstill of the robot activity and need operator interference to recover. The research area addressing this problem is called automated error recovery. Recent research efforts aim at developing systems that autonomously attempt to react to unforeseen situations, either by planning recovery actions, or by applying task rescheduling techniques. An unforeseen situation is not considered an error but merely a situation that requires an adaptation of the planned flow of execution. Unforeseen situations are defined as exceptions, and the corresponding research objectives have evolved from automated error recovery to the more general exception handling.

The nature of the majority of production systems shows an evolution from large batch to small batch or even single products. At the same time, the products get more complicated. So, for smaller batches more sophisticated production tools are needed. These systems bring with them a new problem. This is the problem of dealing with unforeseen situations, errors, and unexpected variations of basic materials and intermediate products. With increasing complexity of the production tools, increasing craftsmanship is needed of an operator to guard the proper functioning of these tools.

In a CIM environment, an attempt is made to automate the craftsmanship of the operator. The task for this automated operator is, in the first place, to handle errors at the operational level of the production process. But also feedback to the design and layout activities is a task for the automated craftsman. Finally, sensor-driven manufacturing can offer a way to overcome the problems associated with small and complex batch production.

The factory of the future might consist of a group of totally autonomous manufacturing cells, each addressing the production of a family of products. Coordination within the cells and among the cells is performed by control systems which are fully capable of planning each operation according to the required production task and the status of the cell components.

However, before such a system is operational, many problems have to be overcome. In the first place, research is needed to develop the techniques needed at the operational level to detect the status of the production system components and consequently react in a appropriate manner. Secondly, the functioning of the operational production system has to be closely linked with the production planning systems. Deviating production times and frequently occurring errors require adjustments in the production plan or even product design. Also at the lower level of production problems have to be overcome. The controllers currently used in industry need to be expanded in terms of computing capacity and communication facilities. Only then will the development of cooperating decision making systems working in real-time response be possible.

Finally, all these new technologies require the specification of adequate standards. In particular the exchange of production information between the various production components and the planning systems needs to be supported by standard information storage and retrieval systems.

The approach taken to realize the above-described situation is to gradually evolve from totally interactive production systems to a fully automatic production environment. In particular, this implies that the various functions of the operator are analyzed and replaced by automated components. In this way subsystems can be tested and applied in existing production environments. Although additional effort is needed to integrate each new development or technique in the existing systems, this approach provides possibilities for industry to benefit directly from new production techniques.

As a step in the evolution of production systems, the ESPRIT 623 consortium studied the problems of exception handling. In this chapter, an analysis is given of the various aspects of exception handling on the operational control level of a robot production system. To study the evolution of manual exception handling to automated exception handling, a well-defined problem was selected. This problem deals with a single robot assembly station and an assembly task.

In Section 17.2, an outline of the exception handling research field is given and the basic notions are introduced. A model for exception handling containing two schemes for exception handling is presented in Section 17.3. A requirement for all

169

exception handling mechanisms is the availability of context information. In Section 17.4, a method is presented for inferring the task's intent from a manipulator level program and a formalism is proposed for representing the global intent of an assembly task. The impact of exception handling on the cell level control is discussed in Section 17.5. An implementation of a robot controller incorporating exception handling mechanisms is described in Section 17.6, and the experimental results of its application are discussed. Finally, in Section 17.7, some open issues and research directions are presented.

17.2 An outline of exception handling

Exception handling capabilities are a basic requirement for a productive manufacturing system. In this section, the underlying ideas of exception handling are presented. In the first place, the aspects of monitoring, diagnosis and exception handling needed at the operational robot control level are outlined. Secondly, an overview of different research results presented in the Chapter is given.

To understand the mechanisms of exception handling, the first question to be addressed is how an operator behaves when dealing with problems and interruptions of the production process. To do so a closer look will be taken on the information which the operator uses and in particular at the structure of the production plan. The objective of planning is to find all the operations needed by the robot to realize a given task and to impose the required constraints on these operations. One of the important constraints is the order in which the operations can be performed. Other constraints are associated with the physical capabilities of the robot and the peripheral equipment needed. Also constraints dealing with the manipulation of the parts can be imposed. A representation of the generated operations and the corresponding constraints is called the plan.

A possible representation for a plan is a precedence graph. In a precedence graph, the subtasks are represented by nodes and the dependency of two subtasks is represented by an arc, connecting the two nodes. The dependency arc indicates that the node which is at the top end of the arc needs to be successfully performed, before the node at the lower end can be executed. In the case of assembly applications, the precedence graph is also called assembly graph.

As a next step, scheduling or sequencing is needed to determine a particular sequence of operations for execution of the plan. Many optimization criteria can be applied to generate this sequence, each depending on characteristics of the robot application. In the case of opportunistic scheduling [6] for example, the sequence of operations is not fixed prior to execution, but the scheduler chooses the next possible operation under the given constraints in such a way that the number of possible second next operations is maximized. Different criteria such as minimization of robot tool change are applied as well.

After the planning and scheduling phases, the result is a sequence of operations for the robot. This sequence is called the robot program. To start production, the robot program is downloaded to the robot controller and executed step by step. During execution of a sequence of operations, exceptions can occur which interrupt the execution order of the operations. State-of-the-art systems often contain safety mechanisms to avoid self-inflicting motions of the robot. Among these are internal limit switches mounted on the actuator axes, power consumption monitoring and a variety of custom-applied switches guarding the interaction of the robot with peripheral devices. If one of these mechanisms triggers, the robot immediately stops all motions and gives an alarm.

At this time the operator is called for. In the first place, he has to check if direct hazard to humans or machines exist. If so, other systems have to be shut down. If there is no further direct danger, the next step is to isolate the erroneous robot and peripheral equipment from the operational production resources. This implies that he has to evaluate which production resources are affected by the exception. In order to do so he must know which task the robot was performing when the exception occurred. This information tells him which other production resources were involved in this task and so he can isolate these as well.

For example, in a parts handling system, several robots are used to transport parts. If a robot is unable to achieve its task of handing over a part to another robot, this other robot will remain idle. Although the latter robot is operational, it is nevertheless unable to achieve its task. It is important to note this, as the second robot could be assigned with a new task.

In the case of an autonomously operating robot, the robot control system needs capabilities replacing those of the human operator. If during the execution of a robot program the actual environmental variables are out of the boundary values, an exception can occur. These exceptions may prevent the robot from achieving its task. To handle these exceptions autonomously, the control system needs three additional functions.

In the first place, the control system must be capable of detecting deviations of environment values from their expected value. In general, tasks have to be carried out by a robot while operating in an uncertain environment. Uncertainty involved in the environment model concerns both the positions of the objects and their tolerances. It derives from factors like (1) the incapacity of perfectly engineering the working environment, and (2) the tolerances in both shape and dimensions of the objects.

The monitoring activity is performed in parallel to the execution of the off-line program. Small deviations of the environment can be handled in a closed loop control. While the program execution proceeds, two additional sources of uncertainty come into play: the inaccuracy of the robot actuators (e.g. during object positioning) and the narrowness of the sensors resolution power (e.g. during object locating).

During monitoring the sensor system is used both for the realization of pre-planned closed control loops and for detecting the occurrence of exceptions. If an exception is detected, control is transferred to the next function.

Once a fault condition is detected, a diagnosis of the current state of the environment is needed to reveal the cause of the fault condition and to classify the exception. For this diagnosis activity, the control system again makes use of the sensor system.

A successful classification of the exception opens up the way for a recovery planning module to plan corrective actions. A correcting strategy is planned and executed, starting from the exception state. The aim of these actions is to restore the environment such that the pre-planned program can be continued.

In order to handle exceptions, the knowledge of the task structure or task intent of the robot program is needed. The information about the task's intent is relevant to the exception handling activity because of the following facts:

1. In order to detect errors during the execution of a program, the actual evolution of the work environment has to be compared with the set of the desired evolutions.

2. In order to plan a correcting strategy, a set of goal states needs to be imposed, that are compatible with the task's intent.

Suitable goal states can be easily deduced from the task's intent. A conventional robot program consists essentially of a sequence of explicit robot oriented instructions or elementary operations. The task structure relates the elementary operations in the program to the task intent. In Figure 17.1, an example is given of the relations between elementary operations and the task knowledge for a handling task.

During the last fifteen years, many research efforts have been directed both at providing robots with self-planning capabilities and at building interpreters for task-level programming languages. In the ESPRIT 623 project, two main attitudes have been shown by the research contributions addressing exception handling in the operational control of robots:

1. Various exception handling methods have been developed, based on the assumption that a suitable task-level description of the robot program is available. Exception handling research focuses on the development of techniques to exploit the degrees of freedom of the robot or the task sequence, to recover from

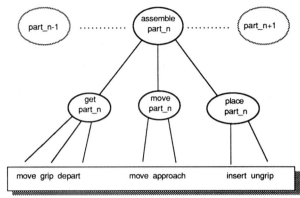

Fig. 17.1   Elementary operations and task knowledge for an assembly
operation.

exceptions and to continue operation. Two mechanisms for exception handling
are described in Section 17.3. These mechanisms are based on the exception
handling model (EHM) [17] which combines application driven heuristics with
general purpose task and path planning functions. This allows, on the one
hand, the problems of applying general purpose planners to specific application
driven domains to be overcome, while on the other hand effective planning
functions can be freely used by the programmer of the robot system. Then the
impact of this model on the explicit programming systems will be explained
(17.3.4).

2. Inferring the task intent from a manipulator level program, some researchers
   have investigated the possibility of extracting as much 'task information' as
   possible from the manipulator level program in order to support the exception
   handling mechanisms. These methods work in particular cases, and some of
   them are able to recognize some 'higher level' actions (e.g. grasping actions,
   carrying actions etc.) by matching simple patterns of instructions within the
   assigned program.

A short analysis of these approaches is presented in Sub-Section 17.4.1. These
methods are not able to extract the global intent of the task, except for some
particular cases. Sub-Section 17.4.2 addresses the problem of extracting the task's
intent from a manipulator level program when :

1. Several desired states can be supported at each program step.
2. Sensing instructions can be prescribed in the assigned program.
3. The program's execution flow can depend on the results of the sensor
   readings.

173

A method is presented for inferring the task's intent from a manipulator level program and a formalism is proposed for representing the global intent of an assembly task. This inference activity can be carried out at compile time, while the collected information about the task's intent can be exploited during the run time exception handling activity.

To reveal the underlying structure of these research areas, the focus will be on an important subset of robot operation, namely assembly operations.

17.3   Two methods for exception handling

17.3.1 Exception handling model

In this sub-section, a model for exception handling is presented. This model is called the Exception Handling Model (EHM). The model consists of two methods of exception handling. Both methods are based on the assumption that a representation of the task structure or task intent is available.

Exception handling is closely associated with general planning problems. In particular when new operations are needed, a planning activity has to be performed. Unfortunately, general planning systems suffer from a complexity problem. If the number of possible operations and the minimum required operations to reach a goal situation are increasing, the needed computation and memory capabilities inhibit an efficient solution. To overcome these problems, two steps can be taken. First, the planning horizon can be reduced, and secondly heuristics can be applied to guide the planning process. The application of these steps to the general planning problem has received considerable attention (eg A* algorithm). What is left as an open question is how a suitable planning horizon reduction and appropriate application heuristics can be determined. In particular, the development of a suitable planning system for exception handling is an open research issue.

EHM addresses the problem of the planning complexity of general-purpose exception handling systems. The underlying assumption of EHM is that the problem of exception handling planning for robotic systems can be reduced by combining application-driven heuristics with general-purpose task and path planning functions. This allows, on the one hand, the problems of applying general purpose planners to specific application driven domains to be overcome, because the heuristic information renders the planning functions superfluous. On the other hand, effective planning functions can still be freely used by the user of the system.

174

Exception handling involves planning and execution of the operations of a robot system which were not foreseen in advance. Therefore, the degrees of freedom which the robot system has within the context of a given task or mission must first be identified. For this purpose, the focus will be on the domain of robot assembly systems, although a similar analysis can be applied to mobile robots. Assembly tasks are often underconstrained and EHM aims at exploiting the inherent degrees of freedom to allow exception handling. The degrees of freedom of a robot assembly task are:
- freedom resulting from the assembly task;
- freedom associated with the operations of the robot system.
Each of these degrees of freedom can be exploited by applying a rescheduling recovery planning technique.

17.3.2 Rescheduling

Any assembly contains parts and possibly subassemblies. The order in which the assembly task can be performed is generally not uniquely defined. Different sequences of assembly are possible, all resulting in the required assembly. In state-of-the-art robot systems, a specific order of assembly is chosen based on some optimization criteria. These criteria can be either associated with the availability of the parts or with the required resources for performing the assembly. Once a particular order is chosen, the robot is instructed to perform the assembly.

If a set S of N operations must be performed to realize an assembly, there are N! possible linear orders of the operations. If the order in which the operations are performed is irrelevant, the plan consists of the collection of the N! sequences. However, for any given assembly task, the possible orderings of the assembly operations are restricted by a set of ordering constraints. The ordering constraints

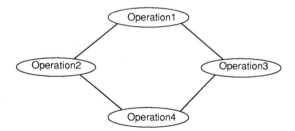

Fig.17.2   Precedence graph example: operations 2 and 3 must be preceded by operation 1. Operation 4 must be preceded by 2 and 3.

175

can be represented by a set C of partial orderings (Oa, Ob), which denotes the temporal relation that operation Oa has to successfully perform before operation Ob. Oa and Ob are elements of the set S which contains all the operations to be performed.

Graphically, the order constraints can be represented by a directed acyclic graph (dag). This graph is called a precedence graph (Figure 17.2). In the precedence graph, a node represents an assembly operation of the robot. An arc between two operation Oa and Ob is called a precedence relation and represents the temporal ordering constraint (Oa, Ob) as defined above.

Given a precedence graph representation of the plan, a linear order of all operations has to be found, obeying the precedence relations. This particular order is called the schedule. As indicated above, if for N operations, no partial orders (in the form of precedence relations) are specified, the schedule has to be selected from the N! possible linear orders. On the other hand, maximum N(N-1)/2 partial orders are needed to reduce the possible linear orders to 1.

The problem of scheduling in robotic applications is to select an optimal order of operations. To solve this problem, the scheduler uses a cost function R over all possible tuples (Oi,Oj) of operations where Oj is the next operation after Oi. The function R can be used to express the costs in term of resources needed or effort required by the robot to perform operation Oj after operation Oi. As such it represents the influence of the robot and its environment on the assembly.

An optimal schedule is now defined as the linear order O for which the total sum of the function R for all tuples (Oi,Oj) is minimal. For example, R can be used to express the costs of tool change. If two operations Oi and Oj need the same tool to be assembled, the value of R is 0. If they require tool change, the value of R is 1. Minimizing RT will now select the schedule with a minimum of tool changes.

During the execution of the assembly task an exception can interrupt the chosen order of assembly. Although a required part of the assembly might be broken down or not available, this does not necessarily mean that no further assembly steps can be taken. The freedom in the assembly order can be used to reschedule the assembly sequence, taking into account that a certain assembly step cannot be performed.

Given a schedule S for realizing the plan:

S: {O1, ...,Oi, ...,On},

a failure of the robot in performing operation Oi will prohibit the execution of operation Oi+1 to On. However, not all operations are affected. Only those operations which directly or indirectly need to be preceded by operation Oi cannot be executed. The remaining operation can still be performed by the robot. To select these operations, all operations that logically follow Oi are analyzed for their dependency on Oi by using the precedence relations C. In this way two new schedules ES and RS are created. ES contains the executable operations and RS the operations for which the successful execution of operation Oi is needed first.

Because the environment of the robot has possibly changed as a result of the exception, the operations of the schedules ES and RS will generally not appear in an optimal order. Therefore, new schedules have to be generated. The scheduling criteria are now associated with the exception that occurred. The cause of the exception and the consequences for a possible alternative sequence have to be taken into account. As these aspects are very much driven by the nature of the application and the use of the resources, a general-purpose rescheduling mechanism needs models of the effect of the exception on the assembly components and the resources. Without knowing the constraints of a particular assembly system, the generation of these models is very complex. This in turn results in time-consuming and complex scheduling mechanisms which do not guarantee that a good solution will be found within the given application environment.

EHM describes mechanisms for the rescheduling of the assembly tasks which embeds the specific application dependent constraints by means of heuristics [14, 24]. These heuristics are specified in so called task rescheduling strategies. The different strategies contain information on how the cost function R has to be adapted to find a schedule with the maximum chance of success. As such, each strategy refers to a distinct scheduling criterion.

In the control system of the robot, the schedule S is now replaced by the schedule ES, and execution is continued with the first operation of ES. After all elements of ES are successfully performed, the remaining elements of the set RS are considered. Of course, the cause of the exception must be dealt with in the first place. Then the elements of RS can be scheduled and executed by the robot.

The mechanism described above can be used recursively. If a second exception takes place a new set RS' is generated. It is merged with the existing RS so that at all time only one set of executable operations and one set of non executable operations is present.

In Figure 17.3 the rescheduling scheme is summarized.

```
- determine if the robot is able to operate
- find a subtask to be executed, not affected by the exception
- start execution of the new subtask and ask operator help to
  handle the failed subtask.
```

Fig. 17.3   Exception handling by rescheduling.

This scheme ensures the continuation of the production process as far as possible, while the restoring of the failed subtask can be performed by a human operator. The default strategy used by the system to find a new subtask is to skip the current subtask and look for the first executable subtasks according to the order in which the tasks are represented in the normal program.

Depending on the application and the configuration of the production system, certain tasks can have a higher priority of being executed than others. For this purpose, a number of rescheduling strategies have been realized which can be used in addition to the default strategy to control the rescheduling process. To apply the proper strategy at the correct moment, the application is modelled in terms of the strategies to follow in case a certain exception occurs.

17.3.3 Recovery planning

In the other approach to exception handling, an attempt is made to recover from the exception by replanning the robot actions. The goal of the replanning activity is to restore the operating conditions in such a way that the off-line generated or normal program can continue. One research direction to tackle this problem is to apply automatic planning systems. These planning systems require a precise description of the environment and a detailed description of the effects on the environment of planning operators. The application of these planners to small scale planning problems has shown the feasibility of the approach. In these cases, the number of objects to manipulate and the number of possible operations is small. The task complexity is therefore low. However, for any robot application of realistic scale, several problems are introduced. In the first place, a suitable representation of the robot and its environment into the planner's world model has to be found. A decision has to be made on which data are relevant to be used by the planner and which are not. In particular if the planning task involves many objects to be manipulated and the complexity is therefore high, a proper selection of all possible operations on these objects is needed. Secondly, the complexity of the task introduces a problem of computing costs. The time needed to compute a possible plan to realize a task increases rapidly if the complexity of the task increases. If the planning problem is not restricted by other means, it can be shown that the complexity of the planning problem is np-complete. In this case, there is no guarantee that a plan can be generated within a given time.

To overcome these problems associated with the complexity of the planning for exception handling, the following requirements must be met:

178

The planning horizon of the planner must be limited. This implies that a limited number of operations is needed to bring the robot system from its initial state into the desired goal state.

Heuristic knowledge concerning the application environment is needed to guide the planning activity. In this way, the planner can be tailored to the application and give a better performance.

The recovery planning scheme of EHM meets these requirements. For the assembly of a part, the robot has to plan a sequence of operations to perform. A typical scenario for an assembly task is 'grasp' and 'assemble' where 'assembly' can be 'place', 'insert', or any other task. This scenario is decomposed into instructions which can be executed by the robot. These instructions are called Elementary Operations (EO). For each elementary operation, geometrical planning has to be performed as well. The result of these planning activities is that a plan is built for the robot to perform the required assembly task. An exception during the execution of one of the operations aborts the execution of all other planned operations. If the robot is still in an operational state, the planning capabilities of the robot can be used to generate a recovery plan.

As pointed out, the occurrence of the exception can have complex consequences for the state of the assembly, the parts, and the resources needed for the assembly. The recovery plan has to take into account these consequences and use an appropriate scenario for the handling of the exception. Just as in the case of the rescheduling mechanisms, in EHM the scenarios are specified as strategies. Each recovery planning strategy consists of a sequence of planning instructions, which together form the recovery plan. Because the definition of the elementary operations of a robot only depends on its capabilities and not on the task or mission, EHM provides a selection of the available recovery planning strategies for each exception that might occur. This so-called kernel already gives the system the capabilities to select proper recovery planning scenarios.

To bring in the heuristic knowledge of a specific application, the kernel of recovery planning strategies can be tailored to the needs of the application. Simply by changing the association between an exception and a strategy, or by altering the order of the strategies, the recovery planning behaviour is adapted. Of course, new strategies can be specified as well. This mechanism for recovery planning was introduced in [14, 19].

As described above, EHM makes use of heuristics to effectively exploit the degrees of freedom in the operation of the robot. On the other hand, EHM uses general functions to transform the heuristics into detailed action plans. For the exploitation of freedom in assembly order, scheduling functions are used. The order constraints of the assembly are represented in adequate formalisms, and scheduling functions are used to sort the possible assembly orders, given criteria corresponding to the required changes of the robot positions and tools. For recovery planning EHM uses functions for task decomposition, path planning, fine motion planning and grasp planning. These functions are general purpose and different variants can be used. Depending on their computational complexity, appropriate functions can be selected by the recovery planner.

Task knowledge plays an important role in EHM. In the first place, EHM uses a formalization of the tasks to determine the extent and the consequences of the exceptions. In the second place, the task knowledge is used to keep track of the context information during the execution of the robot operations [8]. A stack structure is used to store the information concerning the tasks the robot is performing. At any moment, the context of the current robot operation can be retrieved by looking into the stack structure.

Each strategy lists a number of motion or grasp planning activities which are executed one by one. In this way, a detailed action plan is built which is then executed by the robot. The program's task structure is used to find the restarting points for the normal program.

Fig 17. 4 shows the scheme for this exception handling mechanism :

```
- update knowledge of current operating conditions
- classify the exception or exceptions in question
- perform sequence planning by using the predefined
  strategies
- perform motion and grasp planning to build recovery
  plan
- execute recovery plan
- restart normal operation
```

Fig. 17.4   Exception handling by recovery planning.

If no recovery plan can be generated, the goal of the task cannot be reached and the exception handler falls back to the rescheduling scheme of Figure 17.3.

17.3.4 Off-line programming and exception handling

As described in Chapter 14, an off-line programming and simulation environment is needed to create the off-line program. The correctness and functionality of this program can be tested and adapted in simulation. The simulation environment should be transparent to the available robot hardware. This allows the program, when it is sufficiently tested in simulation, to be downloaded to the actual robot controller and executed on-line. To execute the robot program, most current robot controllers run an interpreter which executes the language statements of the robot program one by one. Executing the program the control system makes use of sensor modules which provide the actual value of the environment variables.

In this off-line programming system, it is often very hard to model the work environment of a robot or to predict the value of relevant environmental variables. Many environmental variables have a discrete nature and complex relations with others. As a result, the internal model of the robot environment often only partly reflects the actual status of the environmental variable.

To use a programming language, the off-line programming system must provide a model of the robot itself and its environment [22]. This model contains geometrical data like shapes and positions of objects as well as non-geometrical data. The non-geometrical data represents characteristics of the objects like material type, color, density and information on allowable movements or restrictions [11]. All the modules of the off-line programming system use the information from the internal model. The model should therefore closely correspond to the real situation, to ensure that the resulting robot programs function correctly.

One of the major problems associated with off-line programming is the incompleteness of the internal model of the robot and its environment. In most cases, this model is static and objects are supposed to remain at their location unless they are explicitly moved. In reality, the influences on the environment of the robot are much more complex. Deviations of part sizes, collision with objects not represented in the model or sliding of parts are examples of the influences which could lead to a breakdown of the production process [8, 15]. The conditions of the environment of the robot during the execution of a robot program are called the operating conditions. Unanticipated changes of the operating conditions during run-time of the programs can cause the production system to stop and are called exceptions.

Which programming steps have to be added to the existing off-line programming systems to plan the use of exception handling mechanisms? The aim of EHM is to provide the user of a robot system with a set of strategies which enable the robot control program to recover from exceptions. These strategies must be application independent if they are to function in many different applications. The application-independent exception handling strategies are called the exception handling kernel. To find this kernel, the application-independent components of a robot program must be identified.

In current robot programs, the smallest programmable entities are called the elementary operations. These operations are determined by the capabilities of the robot, but not by the application for which the robot is used. The parameters of the elementary operations as well as the sequence of operations and the program's task structure are determined by the application. The kernel of application-independent exception handling strategies therefore only contains recovery strategies for exceptions which are related to single elementary operations.

181

The specification of the application-independent kernel of exception handling strategies is carried out by a system programmer. Based on expert knowledge of a particular application field like assembly or welding, a set of strategies for possible exceptions is specified. The system programmer identifies the possible exceptions which can occur during the execution of an elementary operation and specifies the handling strategies for all these cases.

The programming model for planning exception handling now consists of two phases. In the first phase, the explicit robot program is expanded with the application-independent strategies for exception handling. In the second programming phase, the application programmer can interactively tailor the available strategies to the needs of the application or add new strategies to the system. This process is similar to the process of robot program simulation, which is used to test and improve the functionality of the robot program. In the same way, the exception handling capabilities of the program can be simulated and improved. The programmer is able to generate exceptions while the robot program is running in simulation. Based on the type of exception, the programmer can trace the flow of decisions of the exception handling mechanisms and check the results.

At this point of programming, the details of the application are known and the application programmer can alter the strategies of the rescheduling scheme and of the recovery scheme. In Chapter 25 the realization of the programming model is described.

17.4 Inferring the task from a manipulator program

Many robot tasks are presently programmed by humans in manipulator languages. Information about the task is heavily impoverished during programming: while a task description has to specify positions, contacts and constraints among the objects, manipulator level programs simply prescribe the instructions needed for task accomplishment. Neither the objects' identities nor the contacts or the constraints among the objects are mentioned in the sentences of a manipulator level program.

The purpose of the method presented in this section is to restore the information about the task's intent, starting from a manipulator level assembly program written in VML. This method exploits a description of the work environment in which both the geometrical and physical characteristics (e.g. shape, dimensions, weight, friction coefficients) of the manipulated objects, and their possible initial layouts are specified. Furthermore, for each part feeder, the occupied space region and the set of the possible objects provided are specified.

The illustration of a proposed formalism for representing the global intent of an assembly task preludes the presentation of the method.

17.4.1 Previous work

In ACRONYM [2], symbolic analysis of uncertainty propagation through a linear plan allows the introduction of additional sensing operations in order to meet the task requirements.

Many errors, such as e.g. those caused by defective objects, cannot be corrected simply by re-executing the last instruction of the program. Therefore it is not sufficient to fully understand the intent of each single instruction. Instead, a representation of the global intent of the program as a whole can enable the synthesis of correcting strategies which follow a completely different path from the one followed by the program.

In [7] the design of an error recovery system is presented, in which some 'task information' is extracted from a linear plan written in a hypothetical object level language. Sensing primitives are not handled, and therefore the employed task representation does not account for multiple expected world states at each program step. This last possibility would allow situations to be represented which are often present in flexible robot programs (e.g. in the presence of part feeders, which can provide objects of several different types).

More recently, some results have been obtained towards the extraction of a task description from an assigned program written in a true manipulator level language (AL) [21]. In this system, a 'preprocessor' presides over the extraction of a task's 'semantics', by using common-sense heuristics in order to come up with the task description that matches some predefined patterns of instructions within the assigned program, (e.g. Open...Move-to-Y...Close_Hand can be matched with GRASP). Some additional information about the manipulated objects can also be extracted (e.g. the grasped object is in position Y). In this way some relationships about the identities of the objects that undergo different manipulations, (e.g. 'equals', or 'differs') can be deduced.

However, the geometrical characteristics of the objects are not exploited. Therefore, the effect of actions that create constraints among objects, cannot always be recognized. Matching patterns of instructions can fail in the following cases: (1) pushed objects, (2) objects with several grasping points, (3) transportation of subassemblies. Furthermore, recognizable patterns of instructions do not include sensing operations. Therefore, the intent of ample parts of the program, including branchings on the results of sensor readings, cannot be fully understood. Hence, global recovering strategies based on the extracted information can fail. In addition, the heuristic rules are only valid for particularly 'gentle' programs.

The determination of the global intent of a manipulator level assembly program is still an open problem. In fact, a systematic approach has not yet been presented, which can be applied to realistic programs. In particular, the following aspects, which are typical of assembly programs, have yet to be considered by the approaches concerning the extraction of the task's intent from a manipulator level program:

1. Several desired states can be supported at each program step.
2. Sensing instructions can be prescribed in the assigned program.
3. The program's execution flow can depend on the results of the sensor readings.

In the following example, the recovery performances of a typical system are shown, in which the above aspects are not considered.

An example:
    Consider the following robot task. The robot has to build assemblies, each consisting of three objects. Object **A** is already present on the work table at the beginning of the execution, and therefore it is not supposed to be defective. The assemblies can be of two different types: (1) **A-B-C**, (2) **A-B'-C'**. The assembly type must depend on the class of the first object, grasped by the robot. Objects are grasped from part feeders. The first part feeder carries objects of two classes (namely **C** or **C'**). Both the other part feeders carry objects of a single class (respectively **B** and **B'**). Object **A** matches geometrically both with **B** and with **B'**. Object **B** matches with **C**; object **B'** matches with **C'**. Neither **B** and **C'** nor **B'** and **C** match together.
    General knowledge about correcting strategies suggests that, if an object is supposed to be defective, then it has to be substituted. If the substitution of an object is not successful, then a further substitution is attempted, of another object belonging to the same assembly. The chronological order of these attempts is based on a 'cost', depending on the physical accessibility of the various potentially defective objects.
    If the insertion of **C** into **A-B** fails, then **C** is substituted with an object **C1**, and the insertion of **C1** into **A-B** is attempted. If it fails, then the substitution of **B** is attempted. Since **A** can not be substituted, since it is present at the beginning of the execution, then, even if the substitution of **B** is not successful, the method is not able to correct the error.
    A description of the task at a first abstraction level can only be extracted if the application of some heuristics rules allows the recognition of some 'macro' level actions. Suppose that, in spite of the problems previously pointed out, the following description of the task can be extracted from the VML program:

```
GRASP(X;Part-feeder-1);
IDENTIFY_CLASS(X);
IF CLASS(X)=CTHEN SET Pos-2=Part-feeder-2
ELSE SET Pos-2=Part-feeder-3;                          */ CLASS(X)=C'
CARRY(X,Pos-3);                                    */ a temporary position
UNGRASP(X);
GRASP(Y,Pos-2);                      */ the second object to be assembled
INSERT(Y,Pos-A);                     */ assembled A and the second object
UNGRASP(Y);
GRASP(X,Pos-3);                      */ the third object to be assembled
INSERT(X,Pos-B);                      */ the assembly is completed
UNGRASP(X);
END.
```

It is supposed that an execution of this program is interrupted because of a defective object, encountered during the insertion of C into A-B. The least 'expensive' strategy, which can be attempted, consists in the substitution of C since it is already held by the robot's gripper. If the new object, grasped from **Part-feeder-1**, belongs to the class C', then its insertion into A-B fails, because of the physical incompatibility among C' and B. This fact leads to the faulty opinion that it was a defect in the object B which caused the error. The subsequent substitution of B cannot remove the incompatibility with C', in that the new object belongs necessarily to the same class B. Therefore, this method is not able to solve this error recovery problem.

Open problems:

Principally, the failure of the above method is due to the lack of information about the global intent of the task: even if insertions can be recognized and distinguished from simple movements (in spite of their similar appearance in a manipulator level version of the program), more global aspects of the task's intent cannot be identified. In particular, both the correspondences C vs. A-B-C and C' vs A-B'-C' are not recognized.

In order to correct the error, the above correspondences among the objects have to be maintained. Therefore, if an object C' results from the substitution of C, then B must be substituted (though not defective) with an object B', before the assembly can be attempted again.

The extraction of the global intent of a realistic manipulator level program suffers some problems that remain to be addressed. These problems are related to the following aspects that can be present in relatively general assembly programs:

185

1. Several states can be foreseen, at a generic program step.
2. Sensing instructions can be prescribed in the program.
3. The program's execution flow may depend on the results of the sensor readings (i.e. it may be unpredictable off-line).

The present approach, the purpose of which is to account for the above 'critical' aspects involved in realistic manipulator level programs, is briefly illustrated in the next section.

17.4.2 Overview of the approach

The purpose of the present approach is the extension of the set of the 'representable' programs in order to include the above aspects of relatively general assembly programs.

A formalism for representing the global intent of an assembly task is first proposed. The main information about the task is supplied by the assigned program. In the proposed formalism, the notion of task involves both the final state of the evolution and some 'external' description of it (the evolving system consists of the robot plus its work environment). In the second component of the notion of task, the effects of the evolution of the system on the external world are included (i.e. the interactions among the external world and the system, during its evolution). Often, the interactions can be schematically subdivided into inputs (e.g. objects or signals entering into the system) and outputs (e.g. objects or signals coming out of the system). In this case, the notion of task can be assimilated to that of the 'behaviour' of the system from an external point of view (i.e. the 'responses' of the system to external 'stimuli').

In a generic assembly program, a set of evolutions of the system can be foreseen. The intent of an assembly program can be represented as the set of the external descriptions (plus the respective final states) of the foreseen evolutions.

A system is presented that extracts the global intent of a manipulator level assembly program. The extraction includes a phase in which the set of the foreseen evolutions is constructed. During this phase, information both about the physical and the geometrical characteristics of the component objects and about the possible initial states of the execution is exploited. The construction of the set of the **foreseen** evolutions is carried out in a way that coincides structurally with a simulation of the program execution.

A difference between the present system and currently available simulating systems derives from the fact that programs are allowed for which: (1) at each step various states of the system are foreseen; (2) sensing instructions are present; (3)

the execution flow may be off-line unpredictable (i.e. the program can not be written as a linear sequence of instructions without branches nor cycles).

17.4.3 The formalism

In this section a formalism for representing the global intent of an assembly task is proposed. The intent of an assembly task can be viewed as the problem to be solved by the robot to which the task was assigned. The assigned program constitutes a partial, algorithmic solution of the problem.

Time is not explicitly concerned in assembly programs written in manipulator level languages. Therefore the following condition, concerning the work environment, must be valid:

Some of the world evolutions, that can derive from the program's execution, are error free: only foreseen events happen during such evolutions. Such evolutions will be referred to as the **desired** evolutions. Once an error occurs, during a program's execution, the desired evolution deriving from the error free part of the program's execution is abandoned. At this point, a correcting strategy has to be planned and executed; the evolution will then consist of an initial desired part and a subsequent real-time planned part. Such an evolution does not belong to the desired ones. However, some world evolutions may exist, which accomplish the task correctly in spite of an error which occurrs (e.g. evolutions deriving from a successful execution of a correcting strategy). Therefore the world evolutions which can be considered **correct** constitute a superset of the desired ones.

A suitable representation of the task's intent must be able to ascertain whether or not a given world evolution is correct. Therefore, the task's intent specifies some requirements, relative to the candidate evolutions. The desired evolutions satisfy these requirements. These requirements must relate to those evolution's characteristics, that are important from the programmer's point of view. In the next subsection the characteristics, concerned by the requirements, are identified.

The I/O Characteristics of An Evolution

The evolving system consists of the robot Space plus its work cell. It will be assumed that the internal evolution of this system is only accessible to the external world (of which the users are part) through some 'communication channels'. These can behave either as inputs or as outputs for the system. The communication channels can be either boxes or tapes, which carry physical objects either towards or out of the system. Therefore, input interactions consist of the entrance of objects into the system through a part feeder. Output interactions consist of the removal of objects (possibly assembled) from the system through boxes or carrying belts.

A generic evolution of the system, deriving from an execution of a program, is described by specifying both the internal evolution of the system and its interactions with the external world. The internal evolution is defined as a discrete sequence of states of the system: the state is 'photographed' at the end of the execution of each instruction of the program (i.e. at each step of the program's execution)

$$Ei = (s0, s1, s2, ..., sn).$$

For each state, both the positions of the objects (including the robot) and their contacts and constraints must be specified. Example:

     ... (position Object1 Position1 State1) ...
     ... (contact Object1 Object2 Surface1 Surface2 State1) ...
     ... (constraint Object1 Object2 Direction_1_2 State1) ...

For each input interaction, both the space region and the classes of the provided objects are specified. Identifiers are added due to implementation reasons.

     ... (input InputChannel1 PickUpRegion)
         (inputResult Id1 InputChannel1 Class1)
         (inputResult Id2 InputChannel1 Class2) ...

Output channels and interactions are described similarly.

In addition, the initial state **s0** of an evolution can be considered as input, while the final state **sn** can be considered as output for the system.

The complexive description of an evolution consists in a sequence of triplets. Each triplet contains one and only one state, a (possibly empty) set of input interactions, and a (possibly empty) set of output interactions:

$Ec =$ (s0,   ({InId_1_1,...InId_h_1},s1,{OutId_1_1,...OutId_k_1}),
         ({InId_1_2,...InId_p_2},s2,{OutId_1_2,...OutId_q_2}),

               ....

         ({InId_1_n,...InId_r_n},sn,{OutId_1_n,...OutId_s_n})).

Unlike the work environments of other tasks (e.g. those accomplished by mobile robots), assembly cells merely constitute sites where some products are processed in order to supply either external machines or human users. Therefore, the details of the internal evolution of the system are of no importance for the user, and hence they are not relevant to the task's intent.

Therefore, the input and output interactions among the system and the external world (including initial and final state) are the only evolution's characteristics, which are important for the task's intent.

The important evolution's characteristics (i.e. the 'I/O' characteristics) are summarized by the following **interaction list.**

Li =   (s0,   ({InId_1_1,...InId_h_1},{OutId_1_1,...OutId_k_1}),
       ({InId_1_2,...InId_p_2},{OutId_1_2,...OutId_q_2}),

                  ....

       ({InId_1_n,...InId_r_n},{OutId_1_n,...OutId_s_n}), sn).

Its first and last elements are respectively the initial state and the final state. The internal elements of the list are pairs of sets (not both empty). Each pair corresponds to a step of the evolution. The order of the pairs, within the list, equals the chronological order of the corresponding evolution's steps. The first set of each pair contains the input interactions, which occur at the corresponding step, while the second set contains the output interactions. The I/O characteristics can be univocally extracted from the complexive evolution: $Li = Li(Ec)$

What Can a Task Require of an Evolution

A suitable task intent is expressed in terms of a set of conditions on a generic evolution of the system. If an evolution satisfies these conditions, then it accomplishes the task correctly. The important characteristics of a system's evolution have been determined: they are summarized in its interaction list. Therefore, the requirements involved in a suitable representation of the task's intent should concern the interaction list of the generic candidate evolution.

Since the desired evolutions are correct by definition, each evolution – whose interaction list coincides with that of a desired evolution – is also correct.

The set T of the interaction lists of the desired evolutions
$$T = \{X/ X=Li(Y) , Y \text{ belongs to } D\}$$
is called 'the interaction set'. If the interaction list of an evolution Ec belongs to T, then this evolution is correct:
$$C = \{X/ Li(X) \text{ belongs to } T\}.$$

The set T of the interaction lists of the desired evolutions is the representation of the task's intent. One can establish whether or not a generic evolution is correct, by ascertaining whether its interaction list belongs to the interaction set.

A more specific formulation can now be given of the error recovery problem: as the program's execution starts, a system's evolution E' derives from it. As the first error occurs, the above (desired) evolution is abandoned. A correcting strategy must be planned and executed. As a consequence of the execution of the correcting strategy, a further evolution E' of the system derives. The correcting strategy must guarantee that the complexive evolution E is correct, i.e. its interaction list belongs to the interaction set.

In order to construct the representation of the task's intent, the set of the desired system's evolutions has to be determined. In the following section, a system is presented that constructs the set of the desired evolutions, starting from both an

189

assigned program written in a manipulator level language, and a model of the environment.

An Example

In this example, the program concerned is introduced into the previous one. The representation of its global intent is shown, and a hint for error recovery is deduced from this representation. At the initial state, the object A is present on the work table. The first action ("grasp X,Pos") prescribes an input interaction. This interaction has two possible results: an object belonging either to the class C (I1) or to the class C' is grasped (I2). Two different desired evolutions can derive from the execution of the program: during the first one an assembly A-B-C is built, while during the second one an assembly A-B'-C' is built. Both the desired evolutions have s0 as their initial state. The input interactions occurring during the first evolution have the following results: (1) I1; (2) I3, i.e. the entry of an object belonging to class B from Part-Feeder-2, into the system. The input interactions occurring during the second evolution have the following results: (1) I2; (2) I4, i.e. the entry of an object belonging to class B' from Part-Feeder-3, into the system. No output interactions occur during the the accomplishment of this task. At the final state of the first desired evolution, the assembly A-B-C is built (state s6). At the final state of the second evolution, the assembly A-B'-C' is built (state s7). The interaction lists associated with the two desired evolutions are, respectively,

$$Li1 = (s0,(I1),(I3),s6) \text{ and } Li2 = (s0,(I2),(I4),s7).$$

Therefore, $T = \{(s0,(I1),(I3),s6);(s0,(I2),(I4),s7)\}$ is the interaction set of the program and represents its global intent.

The method illustrated in the previous example was not able to correct an error caused by a defective object of class C, after its substitution resulted in the entry of an object belonging to a different class C' into the system. The reason for this was that the correcting strategy insisted on trying to build the final assembly A-B-C of the final state s6, while ignoring that no interaction list contains both I2 and s6.

The information about the task's intent represented by the interaction set T suggests how to plan the attainment of the final state s7, in the case where an object of class C' results from the substitution of the defective object (i.e. the result I2 is obtained). In this way the error can be corrected, because I2 and s7 appear in the same element of the interaction set.

17.4.4 Construction of the desired evolutions

The global intent of an assembly task is represented by the interaction set, i.e. the set of the interaction lists associated with the desired evolutions. This section illustrates the system that is in charge of the extraction of the set of the desired evolutions from an assigned manipulator level program.

The system acts as a particular version of a 'simulation' of the assigned program. The simulation starts from the set of the possible initial states of the system. In addition, the system uses both the physical and geometrical characteristics of the relevant objects as well as the space regions through which they enter into the system. In the latter part of this information, the description of the possible results of the input interactions are declared. The simulation can be viewed as the re-iteration of a step. Each simulation step is associated with an instruction of the program. It starts with a set of desired starting states and determines the desired effects of the execution of the current instruction. This results in the construction of the set of the desired arrival states of the simulation step. This set constitutes the set of starting states for the next step. The generic simulation step consists of the (possibly separated) analyses of the **transitions**, promoted by the current instruction, for the various starting states. In this way, for each starting state, the set of the desired arrival states is determined which may derive from the execution of the current instruction. The union of these arrival sets constitutes the set of the arrival states of the current simulation step.

Below, a description of the presented system follows, together with some basic assumptions about the data.

Basic Assumptions and Description of the System

Instructions of the AML language are allowed in the assigned program ( ). Further sensing instructions are also allowed. The assigned program can be represented by a 'program graph'. Its arcs represent the instructions, while its nodes represent the program steps. The program steps are conventionally defined as the sources of the instructions.

A particular execution can be represented as a path of the program graph. A whole set of desired evolutions (hence, not necessarily a single one) can derive from the program's execution. Therefore, at each program step the system's state may vary within a whole set. A program step may also be represented by the associated set of system's states. States are described in terms of **state quantities**. These can be either physical quantities or memory quantities. At each program step, a state quantity may be either constant or variable. During each step of the simulation the execution of a program instruction is simulated.

Actuating instructions can modify (or create) both physical and memory quantities. Computation instructions can only modify memory quantities. Sensing instructions map physical quantities onto memory ones. The test predicates of the control structures depend on the values of some memory quantities. These can depend on the sensor readings.

Exogenous quantities, i.e. both state quantities at the initial step and quantities describing input interactions, may be either constant or variable.

If a program includes a cycle, each program step belonging to the cycle can be gone through more than once by an experiment of execution. It is assumed that during a generic desired evolution of the system, each cycle of the program is iterated a bounded number of times.

Due to this assumption, the original program can be transformed into an acyclic one, by unfolding the program graph.

The transformed program is represented by an acyclic graph. Each step of the transformed program is gone through at most once during an experiment of execution. Each step has a set of father steps, and a set of child steps, that do not overlap.

The simulation results in the construction of an evolution graph, which duplicates the structure of the program graph. Each step is exploded into nodes that represent the desired states associated with the step. Each state is characterized by a set of descriptive quantities and by the value of each quantity. Values can also be logical (like, e.g., for the physical quantity 'contact Obj1 Obj2 Surf1 Surf2").

In addition, to each state s a set of partial interaction lists is associated. Each partial interaction list contains the interactions which occurred prior to reach state s. In the evolution graph, a state s1, belonging to a step p1, is connected to a state s2, belonging to a step p2, if s2 derives from s1 through the instruction that connects p1 to p2 in the program graph. The evolution graph is constructed incrementally during the simulation.

The interaction set of the program, i.e. the representation of the global intent, is constituted by the union of the interaction lists of the final states of the simulation.

In order to be constructed, the evolution graph must have a finite number of nodes. This implies that, at each step of the transformed (i.e. made acyclic) program, either only a finite number of desired states are allowed, or a single node can be used to represent an infinite set of desired states. In order to account for such complex nodes, an extended evolution graph is employed. Actually a generic path, going through a complex node, compactly represents an infinite set of desired evolutions.

Recently the simulator has been extended in order to represent programs providing for infinite sets of evolutions. For details see [1].

17.4.5 Implementation

A system that constructs the set of the desired evolutions of a program has been implemented in KEE on a TI Explorer. This construction is performed by means of a simulation of the possible effects of the execution of the program.

The simulation system is able to track the propagation along the program of the information acquired by sensors such as a camera and a photocell. In particular, the AML instruction 'GUARDI' is simulated,that is used to search for objects by means of, e.g., a photocell. The dependence of the variables describing the subsequent states on the previous position of the searched object is propagated along the program.

A real-time monitoring system is being developed which uses the task information extracted and eventually tries to integrate it with a real-time acquired one. This monitoring system will be employed to monitor the IBM scara robot, programmed in AML.

17.5 Exception handling at cell level

The need for exception handling capabilities in the execution of robot programs is mandatory. In this sense detection, analysis and recovery of errors during execution are phases of the exception handling procedure also to be foreseen by advanced robot cell controllers. The exceptions and their handling phases, which are analogous for other active devices like agv's, machining stations and transportation devices, can have direct consequences for the overall task execution of robot-based manufacturing systems. This section describes the integration of exception handling into production cell task planning (i.e. scheduling) and control. A production cell is a Flexible Manufacturing System (FMS) with several active and passive components devoted to one task. All production systems are subject to disruptive events, ranging from sudden changes in demand to machine failure. Uncertainty is the result of schedule deviations for two general reasons:

(a) equipment malfunction or labour difficulties which results in the later occurrence of events; and
(b) inadequate resources due to fluctuating parts mix, varying batch sizes and late deliveries. Since these events are not predictable in advance, the result is an inevitable need for rescheduling during execution by modifying the existing schedule or generating a new schedule. Schedules are highly susceptible to becoming invalid because errors in the manufacturing process are the rule rather than the exception.

Within a FMS, two control levels are distinguishable, the FMS or cell level and the component or robot level. The generation of robust programs for both levels must include methods for identifying deviant behaviour in the FMS and methods for removal or reduction of the problem behaviour. Feedback information through data collection and monitoring supports detection of unpredictable events, since only the current status of the cell needs to be compared against the predicted state as defined in the schedule.

However, the effects of an unpredicted event can be quite difficult to forecast and difficult to respond to in an optimal manner. For example, a component malfunction might require only rescheduling of the affected subtasks, while a revision of production tasks or the addition or deletion of a critical facility may require a total rescheduling of the cell task. From the point of view of the cell controller, three possible exception handling techniques are available:

- recovery planning;
- (sub-) task rescheduling;
- component rescheduling.

The first two techniques were also used at the single robot level and were explained in Section 17.3. Exception handling by recovery planning is based on the assumption that the environment can be manipulated in such a way that the pre-planned flow of actions can be continued. Recovery planning can take place within various levels of the task decomposition.

In the case of task rescheduling, the exception was due to an error in the parts involved in the production task. All components of the cell are supposed to be fully operational (not affected). Therefore, a selection of the possible operations can be made by inspecting the precedence relations between the parts involved in the manufacturing task.

In addition to these two techniques, a third option is available, which is related to a possible redundancy of production components. This technique is called component rescheduling. Component rescheduling is used to handle exceptions due to a component failure. The component has to be taken out of production and the task has to be scheduled over the remaining components. If an alternative production component can be found to do the job, the corresponding subtasks are scheduled over this component.

No fixed order should be imposed on the application of these techniques. The exception handling mechanisms operating at the component level have to form an integral part with the cell control system. This implies, firstly a correspondence between the control hierarchy and the exception handling levels and, secondly, the integration of the exception handling mechanisms in the underlying model of the cell controller. The requirements imposed by integration and an adequate cell control model representing concurrency and synchronization of the components in the cell were discussed in Chapter 14. In the framework of the presented model the subtask schedules can be expanded with monitoring conditions and rescheduling strategies analogous to the expansion of the robot programs. The first control loop comprises all exceptions to be handled within one subtask without involving the overall schedule. The exception handling strategies go from the reordering of the operations within the subtask to the re-trying of the whole subtask, in case this has implications on the execution of the other subtasks. Now the strategies are strongly dependent on the actual configuration of the cell, i.e., the availability of components having complementary or the same functionality as the component in fault state.

The problem of exception handling at cell level is the subject of many research efforts and has been addressed in the ESPRIT 623 project. Due to the complex nature of cell level exceptions, further work is needed to reveal the possible solutions. One direction currently taken in the CIM-PLATO project is to move away from the strict hierarchical structures on the lower control level and to group the exception handling functions as handling agents controlled by a decision module. Experimentation with industrial test applications is needed to verify these approaches.

17.6   The high level interpreter: a robust robot controller

17.6.1 Overview

The High Level Interpreter (HLI) is the implementation of a robot controller incorporating the exception handling mechanisms of EHM as presented in section 17.3. In this section, the basic components of the High Level Interpreter for exception handling are presented. The programming and simulation of the HLI for exception handling of an assembly task is presented in Chapter 25.

The HLI is not an alternative for the existing robot controllers, but an extension of the existing control functions. To control a mechanical manipulator, the same functions as state-of-the-art robot controllers are needed. These functions, which are dealing with the execution of robot motions, are grouped in a module called the Low Level Interpreter (LLI). The rationale for creating a HLI - LLI is that decision making functions dealing with the flow of execution are separated from the function carrying out the actual execution. In this modular structure, new execution capabilities of the robot mechanics or sensors can be easily introduced without changing the overall control structure. The other way round, new decision making functions or capabilities can be incorporated without affecting the lower level modules. In the next section the functions of the LLI and HLI are described.

Basically, a robot controller consists of an interpreter program, an interpolation program and a hard- or software implementation of a servo algorithm for each robot degree of freedom (joint). The interpreter reads instructions from a file and interprets them. The instructions can be control flow instructions, I/O instructions or motion instructions. Additionally, the instruction set can contain mathematical and logical evaluations. As has been seen in previous chapters, the total of all instructions is called a robot programming language (like VAL or VML) and the group of instructions expressing a task of the robot is called the robot program. If the robot program is generated by an off-line programming system, the robot program is also called the off-line or the pre-planned program.

By stepping through the robot program, the interpreter executes the robot program instructions one by one. For most of the instructions, the interpreter behaves just like a general purpose program interpreter. However, the instruction set also contains motion instructions for the robot. These instructions specify a goal configuration of the robot, a required speed or allowed time and a description of how the motion should be executed. An interpolation program is used to generate the sequence of joint setpoints needed by the servo algorithms. The interpolation program makes use of the interpolation technique specified in the motion instruction (e.g. straight line, point to point, circular) and uses a description of the robot's kinematics and inverse kinematics.

Finally, the servo algorithms control the execution of the motion. A control law is applied which suits the mechanics of the robot best. Often PID control laws are applied. The input of the servo algorithm is the value of the set points and the actual value of the joint position as measured by a sensor (joint encoders). From these input values, a difference value is computed which is transformed in an output value. This output value describes the required torque to be put on the joint and is transformed into a voltage setpoint for the joint motor. The parameters of the control law are preset by the robot manufacturer.

In addition to the functions described above, modern robot controllers are also built to supervise the peripheral equipment of the robot like part feeders, micro switches and conveyor belts. For this purpose, the controller contains a programmable logic circuit (PLC) with a multitude of in- and outputs. Also the controller offers extra servo channels to control an external motor drive (as part of the tool for instance).

The next set of functions concern exception handling. To start with, the robot control system must be capable of detecting deviations of environment conditions from their expected value. This monitoring activity runs parallel to the execution of the off-line generated robot program. The available sensors are thus used for detecting the occurrence of fault conditions.

Once a fault condition is detected, a diagnosis of the current state of the environment is needed to reveal the cause of the fault condition and to classify the exception. For this diagnosis activity, the robot controller again makes use of the available sensor systems.

Finally, a successful classification of the exception opens the way for the recovery planning function to plan corrective actions. The aim of these actions is to restore the environment to such an extent that the pre-planned robot program can be continued.

Summarizing the functions of the robot controller:

- program interpreter and data handling;
- interpolation;
- servoing;
- peripheral control;
- monitoring;
- diagnostics;
- recovery planning.

In the robot control design presented here, the above-mentioned functions have been split into two groups. The functions dealing with the control of the program flow and the supervision of the execution are grouped in the High Level Interpreter (HLI). The HLI contains all the functions dealing with decision making capabilities and the dispatching of motion instructions. The functions needed for execution of motions, peripheral control and processing of sensor data are grouped in the Low Level Interpreter (LLI). The LLI is best associated with a generalization of a modern robot controller: it contains all functions needed for the execution of robot motion instructions. Figure 17.5 represents the model of the HLI and LLI.

The interface between the HLI and the LLI is formed by the motion instructions dispatched by the HLI and executed by the LLI. The motion instructions are called the elementary operations. From this point of view it can be said that the LLI is concerned with all that takes place within the scope of a single elementary operation and that the HLI deals with the execution flow and supervision of the elementary operations.

The off-line planned sequence of elementary operations is called the normal program and when the robot controller is executing the normal program without exceptions, the controller is said to be in normal operation.

197

To represent the task structure in the normal program, the sequence of elementary operations is divided into sections according to the subtask decomposition of the task structure. Before and after each section, additional statements are brought into the program indicating the start or end of the corresponding subtask. An example of the resulting program is given in Chapter 25. In the next table, the elementary operations of a robot arm for assembly operations are listed.

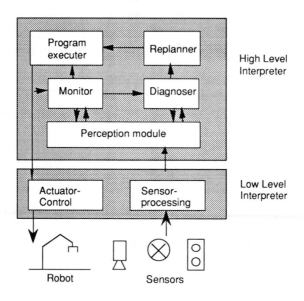

Fig.17.5 High level interpreter and low level interpreter.

Elementary action	Parameters
Transfer	goal, speed
Grasp	part
Detach	part
Insert	goal, forces, speed
Approach	goal, forces, speed
Depart	goal, forces, speed

Table 17.1   Elementary operations.

17.6.2 Monitoring a robot

The robot control system must be capable of detecting deviations of environment variables from their expected value. To realize this, for each elementary operation, a list of monitoring conditions is specified. This list instructs the monitoring module on which environmental variable should be checked when the corresponding elementary operation is executed. The sensor system delivers the values of the environmental variables. An exception is detected by thresholding on the relevant sensor module output. The monitor performs its activity in three phases; before, during and after the execution of the elementary operation. The rationale behind this is that the exception handling strategy to follow depends on the phase in which the exception is detected.

The sensor primitives for an assembly robot system equipped with a wrist force sensor and with a camera placed on the gripper are listed in Table 17.2. A presentation of exception handling within the area of sensor data processing can be found in [25].

Sensor primitive	Description
Check Motion (CM)	Measure movement of robot
Robot Forces (RF)	Measure forces on the robot
Robot Free (RFR)	Measure contact with objects
Plan Position (PP)	Measure current position
Object Available (AO)	Measure availibility of object in gripper
Object Orientation (OO)	Measure orientation of object in gripper
Find Object (LO)	Find location and orientation of object
Identify Object (IO)	Identify object, using data-base

Table 17.2 Sensor primitives.

It is now clear that next to the elementary operations listed in Table 17.1, the sensor primitives of Table 17.2 form an interface between the HLI and LLI.

To monitor the actions of the robot, each elementary operation is associated with a selection of the sensor primitives. For instance for monitoring a transfer operation of the robot, the monitor uses the sensor primitives 'OA' and 'OO' before the operation, 'CM', 'RF', 'OA' and 'OO' during the operation and 'PP', 'OA' and 'OO' after the operation of the robot. In Chapter 25 a full description of the realized monitoring conditions is given.

17.6.3 Diagnostic functions

After detection of an exception, a diagnosis is performed to gain additional information about the nature of the exception and to update the internal model of the robot environment. In general, it is a problem to decide which environment variables need updating and which do not. Fault tree structures are used to guide the diagnosis activity. The input for the diagnosis is the set of sensor module output values which exceeded their boundary values (as detected by the monitor). For each entry in this list, a corresponding entry is given for the fault trees. The fault tree for a specific exception entry consists of a number of arcs and nodes. Each node represents a query to the sensor system to measure the value of an environment variable. Based on the result of the query, a new arc is followed which either leads to another sensor system request, or ends in a leaf of the tree. The leaves of the fault trees represent the possible outcome of the diagnosis.

For each of the elementary operations, a fault tree is modelled. The diagnosis activity results in the classification of the exception. For the assembly robot system studied, five groups of exceptions are identified:

- exceptions resulting from a collision with an obstacle;
- exceptions resulting from the loss of an object;
- exceptions due to calibration and position errors;
- exceptions from extensive force feedback;
- exceptions due to a handling error of an object.

Each type of exception is further divided into several variants, depending on the results of the diagnosis.

17.6.4 The replanner: recovery planning

The replanner of the HLI carries out the actual exception handling task. To do so, it makes use of the two mechanisms described by EHM. As input the replanner uses the exception classification of Table 17.3 and the program status information. The latter consists not only of the identification of the elementary operation which gave rise to the detection of the exception but also of the context information defined by the task structure.

The first attempt of the replanner to handle the exception is to plan a set of recovery operations to bring the status of the robot and the environment back to a state from which the normal program can be continued. This mechanism was presented in section 17.3.2.

The normal program is analyzed to identify the subtask in which the exception occurred. The rationale behind this is that the whole set of elementary operations representing the current subtask will be replaced by a new set of elementary operations which will take care of the new situation. The normal program is searched for from the current position to the statement indicating the end of the current subtask. Now the replanner removes all the elementary operations found. In this way the extent of the recovery plan is kept in limits and the goal position is identified as well.

Next, the new set of elementary operations is planned. The planning activity is guided by the use of application heuristics. These heuristics are called the recovery planning strategies. Each pair of elementary operations/exceptions is associated with a list of strategies. Each strategy itself is sequence of planning instructions. An example of a strategy for handling a lost object is given below:

Strategy: PlanPickUpPart
    sp_findobject(CurrentPart),
    plan_recovery(GetPart(CurrentPart)).

The first planning instruction invokes sensor primitive to identify and locate the current part. The recovery planner uses the task stack to instantiate the CurrentPart variable. The second instruction invokes the task decomposition and planning of a 'GetPart' operation.

As a result of the recovery planning, a new subtask is added to the normal program. This subtask is called 'recovery'. It contains a sequence of elementary operations which handle the exception and fulfil the current subtask. The control is now handed over to the program executor which starts with executing the recovery subtask. After its completion the normal program is taken up again starting with the next subtask.

The recovery planning scheme can be applied recursively. During execution of the recovery subtask, a new exception can be detected, starting the diagnosis and recovery planning over again. The recovery planner keeps track of how many times a recovery plan for the same exception is attempted. Each time an attempt is made, the next recovery planning strategy of the list associated with the elementary operation/exception pair is used. If the list is exhausted, the recovery planning scheme is stopped and the subtask is considered as failed. Now the rescheduling scheme can be applied.

17.6.5 The replanner: rescheduling

In the second attempt of the replanner to handle the exception, the rescheduling scheme is applied. To apply this scheme, the replanner makes use of the precedence relations between the subtasks of the normal program. The idea behind this approach is that although the exception itself can not be handled (the recovery planning failed), this does not necessarily mean that production should be stopped. Therefore, alternative subtasks are selected for execution. During the execution of these subtasks, an external operator (human or machine) is consulted to restore the environment. As a result, the subtasks of the normal program are reshuffled, but no additional planning of elementary operations is carried out.

As described in section 17.3.2, the rescheduler uses a cost function to find a suitable sequence of subtasks. The cost function is also used to express the impact of the exception on the rescheduling activity. To do so, a so-called rescheduling strategy is used. For each subtask/exception pair a strategy is specified which indicates how the cost function has to be changed.

The default strategy used by the system to find a new subtask, is to skip the current subtask and look for the first executable subtasks according to the order in which the tasks are represented in the normal program. An alternative strategy is called 'Skip_master'. This strategy can be used when several tasks are grouped together in a subgraph of the assembly graph (this subgraph is called 'master task'). The strategy then avoids the selection of a task belonging to the same subgraph.

17.6.6 Discussion

The recovery planning scheme used by the replanner is not limited to the lowest level of subtasks of the task decomposition. In principle the recovery planning could take place at all levels of the decomposition. This might be necessary if the recovery planner is not able to find a suitable plan to handle the exception at the lowest subtask level. For instance, a recovery plan could be planned for the 'assemble part' level of Figure 17.1.

In this case the limitations of the planner discussed at the beginning of section 17.3.3, start to play a role. Because the task complexity increases if the subtask level is raised, an adequate recovery plan might not be reached if the planner is allowed to raise the subtask level in case planning for the current subtask level is not possible. To avoid this situation, the application of the replanner must be restricted.

Another reason to restrict the replanner is related to the recursive activation. A possible event resulting in this behaviour is if the actual cause of the exception has not been properly identified by the diagnoser. Each new recovery plan is likely to fail because the real cause of the exception was not properly detected. Finally, the recovery planner stops due to an exhausted strategies list. Clearly, this is not an optimal solution. The fact that the execution of all the recovery plans fails is useful information for the diagnoser. A mechanism is needed to feedback the results of the recovery execution to the diagnoser.

Currently, the diagnoser is also activated each time an exception is detected, but it does not make use of the recovery level information. If this information is used, a better diagnosis could be reached and the recursive use of the replanner restricted. Another interesting option which is possible through replanner/ diagnoser feedback is 'learning by experience'. By keeping a record of detected exceptions and corresponding successful recovery strategies, a more adequate and fast reaction is possible when a similar exception is detected.

17.7 Conclusions and open issues

In this chapter, the incorporation of exception handling into the operational robot control systems was presented. Based on an analysis of the capacities needed to overcome problem situations during robot manufacturing, three additional functions were identified. These are monitoring, diagnosis and planning. The planning function makes use of a model for exception handling. This model contains two schemes for exception handling; recovery planning and subtask rescheduling.

A requirement for all exception handling mechanisms is the availability of context information. This context information is represented by the task decomposition of the robot task. The task decomposition or task intent is used to determine the set of the desired evolutions which are used to detect exceptions when compared with the actual evolution of the work environment. Also, to plan a correcting strategy, a set of goal states needs to be imposed, that are compatible with the task's intent.

A method is presented for inferring the task's intent from a manipulator level program and a formalism is proposed for representing the global intent of an assembly task. This inference activity can be carried out at compile time, while the collected information about the task's intent can be exploited during the run time exception handling activity.

An implementation of a robot controller incorporating the exception handling mechanisms of EHM was presented. The various components of this controller, called the High Level Interpreter, are discussed and the experimental results of its application are discussed. It shows that exception handling functions can be effectively applied to robot assembly systems.

The approach taken in regard to exception handling is to gradually evolve from fully manual exception handling to automatic exception handling. The underlying assumption of this approach is that the knowledge of the application can be formalized and used by the automated system. This knowledge is called the application model and it contains information about the elementary operations, possible sensor systems, exceptions and the task structure. Given a model for exception handling like EHM, application models are needed to build a useful system. So far, application modelling efforts have concentrated on assembly applications. Other application areas like welding, parts handling, free form cutting and mobile robots need to be studied, and suitable application models should be derived.

With the incorporation of exception handling tasks on the operational robot control level, productivity can be increased. However, the current state of the art of robot controllers needs additional capabilities to perform the exception handling tasks. These are:

- extended communication facilities. Interfacing with sensor systems and cell controllers is a necessity to react in time and to adequately detect exceptions. Therefore application level interface specifications are needed on top of a local network;
- additional computing power. The computing models underlying the exception handling mechanisms are based on automated reasoning and knowledge base techniques. The implementation of these models requires sufficient computer memory and sometimes special purpose inference hardware. Also, sufficient programming and debugging facilities are needed. Most general-purpose computer workstations can be configured to meet these requirements. The integration of the workstation approach and a robot controller is needed.

Currently exception handling is applied to the lowest level of the task decomposition. Future robot programming systems will allow a user to program the robot at task level, making use of elementary task primitives. In such a case, the exception handling strategies kernel can be expanded with strategies related to the task primitives of the program.

Ultimately, the factory of the future might consist of a group of totally autonomous manufacturing cells, each addressing the production of a family of products. Programming is carried out by specifying a set of cell task descriptions. Planning, scheduling and dispatching is carried out within the cell controller. Exception handling also takes place entirely within the cell. The exception handling problem is now replaced by the problem of building autonomous systems. The work described in this chapter is a step towards autonomous systems in the shop floor.

References

[1] Bartoletti, M., De Min, S., Mazzini, M., 'Symbolic Simulation of Manipulator Level Programs: Sensorial Instructions' (in Italian) Tesi di Laurea, Politecnico di Milano, 1990.

[2] Brooks, R., 'Symbolic Error Analysis of Robot Planning' Int. Journal of Robotics Research, 1982.

[3] deMello, L.S.H., Sanderson, A, 'And/Or graph representation of assembly plans', National conference on AI 5, 1986, pp.1113.

[4] Donald, B., Error Detection and Recovery, PhD thesis, Springer Verlag 1989.

[5] Ephraim, P., Gaenssmantel, G., 'A Concept of Fault Tolerant Operations of FMS', in: CIM Europ working conference on production systems, May 1986, pp248.

[6] Fox, B.R., Kempf, K.G., 'Opportunistic scheduling for robotic assembly', in: Robotics and industrial engineering; selected readings, Atlanta Georgia 1986.

[7] Gini, G., Gini, M., Somalvico, M., 'Program Abstraction and Error Correction in Intelligent Robots', Proc. 10th ISIR, 1980.

[8] Gini , M., 'Symbolic and Qualitative reasoning for error recovery in robot programs', in: Rembold U, Hörmann K, (eds), Proceedings of the NATO International Advanced Research Workshop on Languages for Sensor Based Control in Robotics, Italy September 1986.

[9] Gini, M., 'The role of knowledge in the architecture of a robust robot control', Proceedings of IEEE 1985, pp. 561.

[10] Henderson and Silcrat, E,. Logical sensor systems. pp. 169 - 193, Journal of Robotic Systems, 1(2) 1984.

[11] Hoffmann, C.M, Hopcroft, J.E., (1987). Simulation of physical systems from geometric models. IEEE Journ. of Robotics and Automation, Vol RA-3, No 3, 1987.

205

[12]   Isermann, R., 'Process fault detection based on modelling and estimation methods-A survey', Automatica, Vol 20 nr 4 pp. 387.

[13]   Malcolm, C., Fothergill, P., 'Some architectural implications of the use of sensors', Proceedings of Intelligent autonomous systems, Amsterdam 1986 pp. 71.

[14]   Meijer, G. R. and Hertzberger, L.O., 'Exception Handling for Robot Manufacturing Process Control', CIM Europe Conference, Madrid, IFS Publications, 1988.

[15]   Meijer, G.R., Duinker, W., Hertzberger, L.O., Kuijpers E.A., Tuijnman, F., 'Monitoring and exception handling for robot control', in: Hirsch, B., Actis-Dato M. (eds), ESPRIT CIM working conference on production systems, Bremen May 1986, pp 235.

[16]   Meijer, G.R., Duinker, W., Hertzberger, L.O., Tuijnman, F., 'Robotbesturing met herkenning en afhandeling van uitzonderings-situaties' (Dutch), CAPE conference, Amsterdam 1987, pp. 610.

[17]   Meijer, G.R., Hertzberger, L.O., Mai, T.L., Gaussens, E., Arlabosse, F.,'Exception handling system for autonomous robots based on PES', Proceedings of Conference on Intelligent Autonomous Systems 2 (IAS-2), Amsterdam 11-14 December 1989, Elsevier Science Publishers.

[18]   Meijer, G.R., Weller, G.A., Groen F.C.A., Hertzberger, L.O., , 'Sensor based control for autonomous robots", Proceedings of IEEE international conference on control and applications, Jeruzalem April 1989, WP-3-4.

[19]   Meijer, G.R, Hertzberger, L.O., 'Off-line programming of exception handling strategies'. Proceedings of IFAC/SYROCO conference, Karlsruhe October 1988, VDE/VDI publications.

[20]   Rault, A., Jaume, D., Verge, M., 'Industrial process Fault detection and localization', Proceedings of IFAC 9, Budapest Hungary, 1984.

[21]   Smith, R., Gini M., 'Robot Tracking and Control Issues in an Intelligent Error Recovery System', Proc. 1986 IEEE Conference on Robotics and Automation, San Francisco, 1986.

[22]   Stobart, R.K. (1987), 'Geometric tools for the off-line programming of robots'. Robotica Vol 5, 273.

[23]   Taylor, R.H., 'An integrated robot system architecture', Proceedings of the IEEE, Vol 71 no 7, July 1983, pp.842.

[24]   Tuijnman, F., Meijer, G.R., Hertzberger, L.O., (1989), 'Data modelling for a Robot Query Language', Proceedings of Conference on Intelligent Autonomous Systems 2 (IAS-2), Amsterdam 11-14 December 1989, Elsevier Science Publishers.

[25]   Weller, G.A., Groen, F.C.A., Hertzberger, L.O., (1989), 'A sensor processing model incorporating error detection and recovery', Proceedings of NATO advanced research workshop on Traditional and non-traditional sensors, Maratea 28 August - 1 September 1989, NATO ASI series.

[26]   Wonderen,G.M. van. (1988), AI and high level robot control and a preliminary implementation of a High Level Interpreter. Technical report, Faculty of Mathematics & Computer Science, department of Computer Systems, July 1988.

Chapter 18

Conclusions

K. Hörmann
University of Karlsruhe, Germany

Powerful and versatile robot programming methods are of essential importance for integrated systems. Research and development in the ESPRIT 623 Project has focussed on two important aspects: the realization of an explicit (motion-oriented) programming and simulation system, and the realization of an implicit (task-oriented) programming system. This is the subject of Section III. Using software modules from various international project partners, demonstrator systems have been constructed to test and demonstrate the methods with several industrial applications. Using these new techniques, high time savings (between thirty and eighty-five percent) can be achieved.

An important issue in this context is the generation of executable trajectories. For this purpose, a homogeneous operation environment has been developed featuring tools for motion programming, analysis, calculation, and optimization (Chapter 12). Thereby, powerful graphical realtime animation has been employed. The results are smooth trajectories with minimal execution time and minimal conture errors.

When moving with an application to a new type of robot, it is desirable to reuse the old robot program. This, however, is normally not possible because of the language incompatabilities of different controllers. For this purpose, in ESPRIT 623 an intermediate code called ESR (Explicit Solution Representation) has been defined (Chapter 13). This code can be generated by compilers or task-level planning systems. Using this code, a program generation module is able to generate executable code for various robot controllers.

To simplify offline programming and to shorten the time needed for program development, graphical simulation systems are very important. A process execution simulation system has been developed which allows robot programs to be tested in order to verify whether their manipulation tasks will succeed or not (Chapter 14). Programs, creating off-line can also be optimized and can be used for the planning and testing of workcell layouts. This significantly speeds up the realization of new applications.

When implementing a robot program, the problem arises of how to adapt the off-line-generated progam to the actual working environment. These problems are due to small deviations between the model of the workcell and the reality and are also due to the positioning accuracy of the robot. For this purpose, powerful robot calibration procedures have been developed. Using error compensation methods, the offline-generated programs are corrected (Chapter 15).

A task-level robot programming system has been developed which is able to automatically find the sequence of actions in order to solve the specified assembly task. This involves complex geometrical methods to analyse the target assembly task. This results in precedence constraints between single assembly operations. These operations are assigned to different machines using scheduling methods. In order to generate safe trajectories, collision-free path planning is used. Also, automatic grasp-planning methods have been developed which allow the generation of collision-free grasping motions for a three-fingered hand (Chapter 16).

Finally, unexpected situations often may arise during the actual execution of robot programs. These can lead to a complete standstill of the workcell and normally require operator interference. Substantial research has been undertaken to develop systems which are able to autonomously react to unforeseen situations. This is done by either planning recovery actions or by applying task rescheduling techniques (Chapter 17).

209

Section IV

Information system

Chapter 19

Introduction to the information system

F. Sastron
Universidad Politecnica de Madrid, Spain

19.1   The process of planning a flexible manufacturing system and its support

Planning a flexible manufacturing system is a complex process. It receives information from previous activities about the product to be manufactured and the amounts to be manufactured per time unit. With that information, a manufacturing system able to solve the posed manufacturing problem has to be designed. This implies two basic activities: planning the system layout and planning the execution process. Both planning activities are usually performed by stepwise refinement and are closely interrelated.

The number of computer-based tools that support parts of the planning process has grown very much over the last years. Most of the tools in this project and others outside it belong to the category of planning tools.

'Information System' is an ambiguous term that refers to a different set of tools which give support to information integration activities. Information integration here means the building, out of a certain number of planning tools, of a coherent system which helps the planner of a flexible manufacturing system to travel through the planning life-cycle in an integrated way. This is the subject of Section IV.

When accomplishing the aforementioned building task, some important problems emerge. Among them, are the following:

1 Tools have to be able to communicate with other tools, even when residing on different computers or running under different operating systems.
2 The users have to be able to communicate with the tools in a homogeneous way.
3 Integration from the logical point of view is based on building global models in which the relevant aspects of the planning process and the relationships between them are properly captured. The complexity of this modelling task calls for very expressive modelling techniques. Standardization of these techniques is needed to allow information exchange.
4 Planning a flexible manufacturing system is a process which belongs to the more general family of design processes or processes in which engineering artifacts are devised. Design processes have a number of specific features which do not appear in business applications and which need proper administration with the corresponding computer support.

19.2   Approach adopted

The four problems just mentioned are at present the subject of active research, development and standardization works. Among these are:

- Some European projects belonging to ESPRIT and EUREKA;
- Some projects financed by the DoD of the USA;
- ISO TC 184 "Industrial Automation Systems" and ISO/IEC JTC1 "Information Technology".

Many of these initiatives concentrate on creating a basic infrastructure for information integration with different points of view which range from the very general integration problem to the problems of integration in the engineering field or the integration problems in CIM.

Work in this project has mainly focused on problems three and four of the previous section. The main topics have been: modelling of the complex entities and relationships which appear in planning flexible manufacturing systems, integration of information from multiple sources (CAD system, relational database, cell controller, etc.); and management of the planning process.

Two complementary approaches have been followed, knowledge- and data-based:

- A hybrid approach based on a Knowledge Base Management System (KBMS) as a modelling kernel and integration front end for multiple types of information sources.

KBMSs have powerful information representation capabilities which make them ideally suited for building the type of complex model that appears in design or engineering activities. They also incorporate various programming paradigms which allow them to act as an integration kernel for information coming from multiple types of sources.

An approach that uses a relational Data Base Management System (DBMS) as a kernel to build a kind of engineering database called design management system.

DBMSs have been the basic element for the solution of information integration problems for decades. Today, relational DBMSs are a stable technology with an increasing acceptance due to their flexibility and data independence.

These two approaches are now presented in Chapters 20 and 21. Chapter 22 summarizes and concludes this section.

Chapter 20

Design management system

F. Sastron, A. Jimenez
Universidad Politecnica de Madrid, Spain

20.1   Introduction

The term Design Management System (DMS) refers to a system that supports a determinate design process, integrating the data management aspects of the various tools that cooperate in the design activity. The design process considered in ESPRIT Project 623 has been the planning of robotic assembly workcells.

In the area of conventional or business applications, data base management systems have been the basic element for the solution of information integration problems during decades. Today, products based on the relational data model seem to have the widest acceptance due to greater flexibility and data independence.

The relational data model of Codd appeared in 1970. During the early eighties it began to be used, mainly by the electronics design community, to manage design information [2, 4]. Now the number of CAD tools in mechanical design has also multiplied and similar problems are arising.

In this chapter a prototype DMS will be proposed to support the workcell design process which has as its kernel element a relational data base management system (RDBMS).

20.2   Elements for a DMS

Complex design processes are usually composed of a certain number of design steps (DSs). Design is mainly carried out at workstations connected in a network, by designers that normally specialize in one or a few DSs.

The work of a designer performing one DS begins by extracting from a central DMS a part of the design, usually a set of representations of technical objects. From these representations other representations will be derived which advance one step in the design process. In this derivation the designer will use his own domain knowledge as well as auxiliary information given by the computer, for example, information about manufacturing devices in the case of workcell design. This is illustrated by Fig. 20.1, where DINn means design representations input to step n, DLn means design level or set of design representations output from step n, and AUXn means auxiliary information given to step n. Double line arrows mean data flow and single line arrows mean control flow.

The different DSs that make up a design process are related by precedence relationships. Precedence relationships between design steps can be expressed by means of a directed graph whose nodes are design steps and whose edges indicate that the DS at the origin has to be completed before the DS at the end may start. This graph will be callled a precedence graph of design steps. Fig. 20.2 gives an example.

In the case of workcell planning, this graph is simply a list of design steps. This has been given by the reference model of the Systems Planning subgroup, within which our the DMS has been developed.Future development of the aforementioned reference model may lead to more complex structures, like the

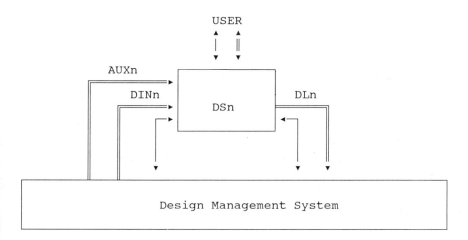

Fig. 20.1   Relationship between the DMS and a DS.

217

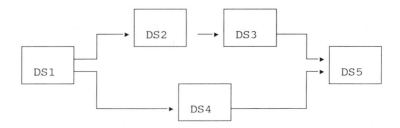

Fig. 20.2   Precedence graph of design steps.

ones represented in a directed graph. For this reason the prototype presented here has been developed in such a way that it can grow easily in that direction.

Workcell planning as it stands now in the project has the following design steps:

DS1 - Obtains the precedence graph of the assembly operations, called assembly sequence, and candidate devices to perform the assembly operations, based on the initial specification of the assembly problem;

DS2 - Obtains the cell layout and simulates the corresponding material flow;

DS3 - Obtains the assembly process in which the operations of the assembly sequence are detailed and linked to the equipment that will perform them.

20.2.1   Three types of data representations

The data representations that appear in DMS can be divided into three groups:

1 Representations of the technical entities which are created in the design process like assembly sequences or cell layouts. They will be called design objects (DOs). In Fig. 20.1, DINn and DLn are composed of one or several DOs. In the present approach to workcell planning, the following design objects have been considered, grouped by DL:

DL0
        DO0.1   -   Initial specification

DL1
        DO1.1   -   Assembly sequence

218

DO.1.2   -   Candidate devices

DL2

DO2.1   -   Cell layout
DO2.2   -   Results of materials flow simulation

DL3

DO3.1   -   Assembly process

In this list, DL0 is not the output of any design step inside workcell planning but is the external initial specification.

2   Representations of the technical entities that are common to all designs and used as auxiliary information by all DSs, like robots or feeders. They will be called catalogue data representations. In Fig. 20.1, AUXn contains data obtained from the catalogue.
In the current approach, the catalogue contains information about the manufacturing devices used in assembly: feeders, fixtures, robots, sensors, storage devices, tools and transports.

3   Representations of the entities that allow the management of the design process, called management data representations. The following paragraph explains them.

20.2.2   Management entities

Management is based on the entity called 'version'. A version describes an instance DL. That is, when a DS is successfully executed, generating one or more instance DOs, a version is created and the identifiers of the DOs are stored in it. Every version is connected to the version of the precedent DL in the DL list, which allows the evolution of the instance design process to be reconstructed from its beginning to its present state. The corresponding data structure is a list of versions.

Consistency and concurrency are managed by means of indicators kept in versions. Other management data like dates and people involved are also kept.

Design is an iterative and tentative activity. A designer who is performing DSn, based on the precedent DLs of that particular design, will normally find several alternative solutions depending on the strategy he adopts. These solutions need to be recorded until an evaluation step is performed to decide which alternative will be selected. The corresponding data structure is a tree of versions.

An example of a version tree can be seen in Fig. 20.3. Versions are depicted as rectangles. Several of the implemented attributes of a version are shown. V_ID means version identifier. DO_ID means design level and CONS_STATE means consistency state. Identifiers are shown as integers of length 4 for clarity; in the implementation they are strings of length 8. READ_LOCK and WRITE_LOCK are indicators for concurrency management.

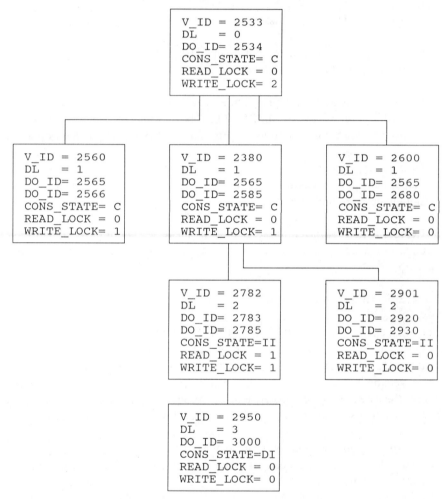

Fig. 20.3   Version tree.

220

This version tree could correspond to a workcell planning that has progressed up to DS3 and is possibly not complete. The first version in the tree has V_ID=2533. From this initial specification, three alternative versions of the next DL have been derived. All three have the same first DO_ID = 2565 which means that the assembly sequence is the same for the three alternatives, while the candidate devices are different in each of them. The design has been continued following the second alternative; DS2 has been executed twice, deriving two alternative cell layouts and the corresponding materials flow simulation results. DS3 has been applied to the first alternative deriving a new version of the first alternative which then derives a new version of DL3 which is associated to the assembly process with DO_ID = 3000.

20.2.3 Manipulation of version trees

Application programs included in the DSs manipulate version trees by calling procedures and functions that perform the manipulation operations. The basic ones are the following:

MAKE_NEW_VERSION: The DMS is requested to create a new node of a version tree i.e. a new version.

OPEN_VERSION: The DMS is requested to access a version in one of three possible access modes: RD (to read the associated DOs), WR (to modify the DO associations of the version) or DL(to delete the version and its successors, including the associated objects if they do not belong to any other version).

Depending on the consistency state of the version and the read or write locks acquired by other designers on it, the access is granted or denied.

If the access is granted the corresponding locks are set on this version and related versions to restrict future OPEN_VERSION requests on this version tree, managing in this way concurrent accesses. If the access to the version is granted, the designer is allowed to call the DO manipulation procedures which correspond to the ACCESS_MODE requested. They will be presented in paragraph 20.2.4.

CLOSE_VERSION: The DMS is requested to complete the access that was initiated with an OPEN_VERSION request, in one of two possible finish modes: commit or abort. In both modes locks acquired at open time are removed and the consistency state of this and related versions in the version tree is updated.

The sequence of operations contained between an OPEN_VERSION and its corresponding CLOSE_VERSION will be called 'Design transaction'. Design transactions are the unit of consistency. This type of transaction differs from conventional transactions in that it is conversational, as indicated in [1]. This makes design transactions much longer than conventional ones and therefore they are not good as recovery units, as explained in paragraph 20.3.

A designer may have any number of opened versions at any moment. One of them will be called the active version. When the designer opens a version, that version is taken as the active version. The active version can be assigned within the list of versions opened by the designer.

A design transaction may be interrupted and resumed later on.

Some other routines useful for the manipulation of the version trees are available.

20.2.3.1 Consistency management

A type of global consistency management has been implemented using the version tree. The ideas presented in [4] have been applied to this case where the semantics of the version concept is a little bit different.

When a version is created, the DMS sets its CONS_STATE=C, i.e. the version is marked as globally consistent, which means that the designer has derived it from its predecessors in the version tree.

When a version is modified, it is understood that its successors in the version tree need modification in order to make them consistent with their modified predecessor. For this reason, when a CLOSE_VERSION with FINISH_MODE=COMMIT is issued after modifying a version, its successors are marked as globally inconsistent. Versions on DL below the modified version are marked as independently inconsistent, CONS_STATE = II. Versions more than one DL below the modified version are marked as dependently inconsistent, CON_STATE = DI, because they cannot be updated until they reach the II state once all predecessors have reached the C state.

For example, in Fig. 20.3 a modification was done in one of the design objects of the version with V_ID = 2380. After closing the version with commit finish mode, versions with identifiers 2782 and 2901 were marked as II because they are not supposed to correspond any more to their predecessors. The version with identifier 2950 was marked as DI, and therefore the DMS will not allow any designer to open it in write access mode until version 2782 gets the consistent state.

The following global consistency rules have been implemented in DMS:

- No version can be derived from an inconsistent version. Implemented as a precondition of MAKE_NEW_VERSION;
- Any version can be read;

222

- A version can be updated only if it is consistent or independently inconsistent. Implemented as a precondition on OPEN_VERSION ACCESS_MODE = WR;
- Any version can be deleted. If a version is deleted then its subtree will be deleted. Implemented as a precondition and an action of OPEN_VERSION ACCESS_MODE = DL.

20.2.3.2 Concurrency management

Management of designers' concurrency on a version tree has been implemented into DMS [4].

Concurrency accesses on the same version follow the usual rule that only read accesses are compatible, any other combination of access modes between the present and the requested OPEN_VERSION being forbidden.

When one designer has been granted access to a version, DMS will restrict the access to its predecessors and successors in the version tree in order to avoid that the access of a second designer on a related version invalidates the work of the first designer.

If a version is being accessed for read, the version itself and its supertree (set of predecessor versions) will remain stable, i.e., no new write or delete accesses will be granted on them. This is achieved by means of the WRITE_LOCK integer attribute that belongs to every version. When this variable is set to a value other than zero no new write or delete access to the version will be granted. When a DS is granted a read access to a version, the WRITE_LOCK attribute of the version and of all the versions in its supertree is incremented by one.

For example, in Fig. 20.3 a read access to version 2560 was granted to a DS and in consequence its WRITE_LOCK and the one of version 2533 were incremented by one.

If a version is being accessed for write or delete, the version itself and its supertree will remain stable by setting the corresponding write locks as explained before. The version itself and its subtree (set of successor versions) will not be visible to other designers, i.e. no read access to it will be allowed. This is achieved by setting the READ_LOCK of the version to one. When a read access on any version is attempted the whole supertree is examined and, if any version in it is read-locked, the access is denied.

For example, in Fig. 20.3 a write access to version 2782 was granted and in consequence its WRITE_LOCK and the one of versions 2380 and 2533 were incremented by one. Moreover, the READ_LOCK of version 2782 was set to one preventing in this way read access to it and to version 2950.

223

		Predecessor		Same Version		Successor	
		RD	WR or DL	RD	WR or DL	RD	WR or DL
Present access on version	RD	Y	N	Y	N	Y	Y
	WR or DL	Y	N	N	N	N	N

Fig. 20.4   Concurrency rules.

The rules that govern concurrency are presented in Fig. 20.4, where Y means that the concurrency precondition to grant the access is satisfied and N means the opposite. In the Y cases the consistency preconditions still have to be satisfied for the OPEN_VERSION to succeed. In the N cases the OPEN_VERSION fails.

At the end of the design transaction, when the CLOSE _VERSION is issued, the locks set by the corresponding OPEN_VERSION are removed.

20.2.4   Manipulation of design objects

Procedures to manipulate DOs are called inside a design transaction. They are as follows:

GET_DO: The DO corresponding to a given identifier which must be associated to the active version is read from DMS and passed to the calling DS.

COPY_DO: Associates a DO that is already stored in DMS to the active version.

PUT_DO: The new DO created by the DS is given an identifier, it is associated to the version and it is stored in DMS.

ERASE_DO: The DO indicated is disassociated from the active version.

In Fig. 20.5 an example is shown of a design transaction in which object 2920 (a cell layout) of version 2901 in Fig 20.3 is updated, creating a new object that is associated to the version. Only the basic elements are presented:

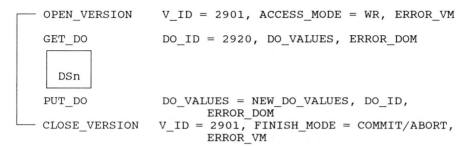

OPEN_VERSION	V_ID = 2901, ACCESS_MODE = WR, ERROR_VM
GET_DO	DO_ID = 2920, DO_VALUES, ERROR_DOM
DSn	
PUT_DO	DO_VALUES = NEW_DO_VALUES, DO_ID, ERROR_DOM
CLOSE_VERSION	V_ID = 2901, FINISH_MODE = COMMIT/ABORT, ERROR_VM

Fig. 20.5 Example of a design transaction.

20.3 Overall structure of DMS

The overall structure of DMS is presented in Fig. 20.6.

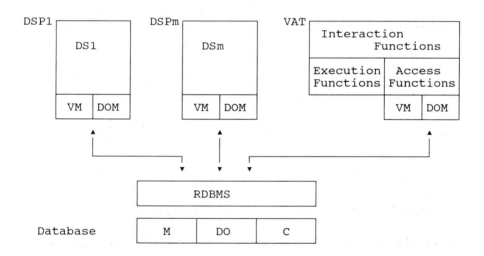

Fig. 20.6 DMS structure.

DMS is built around a conventional relational data base management system. The relational data base contains data representations of the three types used by DMS, presented in previous paragraphs:

- management data representations (M);
- design objects (DO);
- catalogue data representations (C).

A design step program (DSP) is composed of two parts:

- The code corresponding to the design step (DSn) itself, including calls to the procedures for version and design object manipulation;
- The procedures for version manipulation (VM) and design object manipulation (DOM).

The last object presented in Fig. 20.6 is the version access tool (VAT), an interactive interface to DMS that will be presented in paragraph 20.3.1.

Every DSP contains one or several design transactions. A single DSP could contain only one design transaction, like the one in Fig. 20.5.

Every VM or DOM procedure basically consists of a conventional transaction that accesses the data base and performs the high-level management algorithms outlined in previous paragraphs.

A DSP therefore spans a certain number of design transactions and each design transaction spans in turn a certain number of conventional transactions. This is illustrated in Fig. 20.7 for a DSP with a single design transaction.

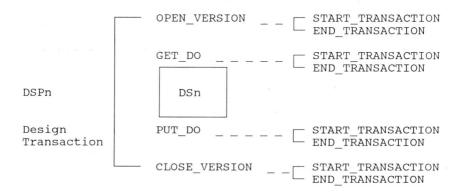

Fig. 20.7 Conventional transactions inside a design transaction.

If DSn is a very complex application that is executed for days or weeks then recovery in case of a failure when DSn is executing could be a problem since a lot of work would be lost. One approach to improve recovery in such cases is to revise the list of design steps splitting those that are very long, until every DSn has a size in correspondence with the maximum amount of work one allows the system to lose in case of failure.

Additional programs not shown in Fig. 20.6 allow the catalogue to be accessed and its maintenance to be performed.

20.3.1   The interactive interface

The module called VAT is an interactive interface to DMS which runs on a graphics display. It is composed of four parts:

- Interaction functions that control the interaction with the user, based on menus, windows, graphics (2D and 3D) and pointing devices;
- Access functions that allow the user to browse through designs. The following are available:
    - Display of a version tree in 2D showing basic attributes of versions;
    - Display design transactions in execution by the user, i.e. versions by him opened and not closed, marking the active one;
    - Interactively issue VM calls;
    - Interactively issue DOM calls. Creation of design objects is reserved to the DSs and therefore PUT_DO is not included in VAT;
    - Display geometry. Geometry is not stored in the data base; instead, names of CAD-files are stored. For parts and assemblies the geometry can be displayed.
- Execution functions that allow the user to start the execution of a DSP passing to it the identifiers of the required versions;
- Procedures VM and DOM.

20.4   Conclusions

The system presented in this chapter attempts to support the management of the planning of robotic assembly workcells in the framework of the management of design processes. Data structures are adopted which abstract high-level characteristics of the design activity e.g. being stepwise and tentative. Restrictions are placed on the access to those data structures in order to enforce high-level or global consistency.

The system is built around a commercial relational data base management system which stores the management entities, the objects generated by the design steps and the catalogue entities. High-level consistency is programmed into the access procedures through which the application programs access the information.

DMS attempts to provide the programmers of workcell design applications with a high-level interface which supplies them the aforementioned functions for global design management and the technical entities they need in an adequate high-level format which hides from them the technicalities of relational data base design. The next chapter describes a complementary approach to information integration, based on a knowledge base management system.

References

[1]   Haskin, Lorie, *On Extending the Functions of a Relational Database System*, SIGMOD Int. Conference on Management of Data, 1982, pp 207-212.

[2]   Katz, *Managing the Chip Design Database*, Computer, Dec. 1983, pp 26-36.

[3]   Lorie, Plouffe, *Complex Objects and their Use in Design Transactions*, ACM-IEEE Database Week, San José, 1983, pp 115-121.

[4]   Neumann, Hornung, *Consistency and Transactions in CAD Database*, 8th Int. Conf. VLDB, Mexico City, 1982, pp 181-188.

[5]   Neumann, *On Representing the Design Information in a Common Database*, ACM-IEEE Database Week, San José, 1983, pp 81-86.

Chapter 21

Design of a knowledge-based information system

Universidade Nova de Lisboa, Portugal
A. Steiger-Garção, L. Camarinha-Matos

21.1 Justification of the approach

This chapter presents an approach to the realization of an information system (IS), which aims at a greater degree of integration by combining different technologies (KBS, DBMS; CAD), and thus opening the way for more ambitious control systems. In this approach, developed by UNL, a Knowledge and Information System (KIS) is the kernel for integration of all functional modules of the planning and execution monitoring system.

21.1.1 Goals for integration

The development of advanced manufacturing systems bring up two important requirements: flexibility and cooperative decision making [16] [8].

Flexibility, or its synonym adaptability, is here seen in terms of the system's capability to cope with changes at all stages of the production process, namely changes in product requirements, changes in technologies and processes, surprises on predicted effects and a capacity to recover from such exceptions. Another view of flexibility is related to fast setup times, in the sense of reduction of programming effort.

Cooperative decision making is a requirement because, even though a progressive automation of functions is desired, the user should interact with the system and his role must be clearly stated in the system's design phase [2][3]. In fact, interactive approaches, mainly based on graphical interfaces, are becoming available. An integration of automatic decision making functions with interactive cooperation of human experts is now a realistic challenge.

Flexibility implies a multi-level feedback mechanism, that is, a capability to acquire real data, 'compare' it against expected models and generate recovery procedures to compensate deviations. This decision making has to have access to the information and knowledge used/produced during design and planning phases. On the execution monitoring level, the IS can be a natural way for supporting the instantiation of values that are accessible only in run time (actual position of a part, for instance) and 'conveying' such information to subsequent operations. Additionally, the IS will 'assist' the monitoring activity by providing the system's expectancies and the knowledge for diagnosing and recovering exception conditions [11]. Therefore, a knowledge and information system constitutes the integration link between planning and execution monitoring.

In the same direction, cooperative decision making, where multiple human experts can incrementally and asynchronously provide contributions to problem solving, has to be supported by such a knowledge and information system that will 'accumulate' all contributions (in the way of a structured blackboard) and support coherence checking.

Under a CIM framework, several almost specialized activities have to be considered together [12], [14]. However, different backgrounds are in general associated with them leading, up to now, to strong difficulties when attempting their integrations. Moreover, as some of these activities are strongly interdependent, it becomes extremely difficult to close any required feedback loop.

For instance, most of the work performed at the level of product design, process planning and cell layout is related to programming and simulation in the sense that generated information is vital for these latter activities and some eventual feedback can impose modifications on design and planning. One can say that design and planning activities represent a preparatory phase of vital importance for subsequent programming and execution supervision tasks. On the other hand, in order to evaluate the feasibility and performance of a proposed solution, the production plan has to be conceived and probably simulated for a given cell, which shows how some feedback from the technological process could lead design engineers to find alternative/refined solutions for the product/cell, better suited to the proposed production process (and vice versa) [13] [7].

At the execution supervision stage, a diagnosis capability requires knowledge about the intended purpose for each action and cell capabilities. Planning of recovery actions also implies knowledge about the actual cell and process execution status as well as basic planning knowledge.

However, separate developments and the nonexistence of common (shareable) information support makes an effective realisation of the required interactions very difficult. In fact it can be stated that a functional integration is not viable due to the use of multi-vendor components, with non-standardised interfaces, operating at different abstraction levels and showing different degrees of automation; however, an integration of information provides an effective way of allowing each activity/function to take advantage of the results generated by other activities. Therefore the Information Systems (IS) concept is, to our view, a solution to the integration of these multiple activity areas [2][6][15] and for 'recovering' of existing tools (see Fig. 21.1).

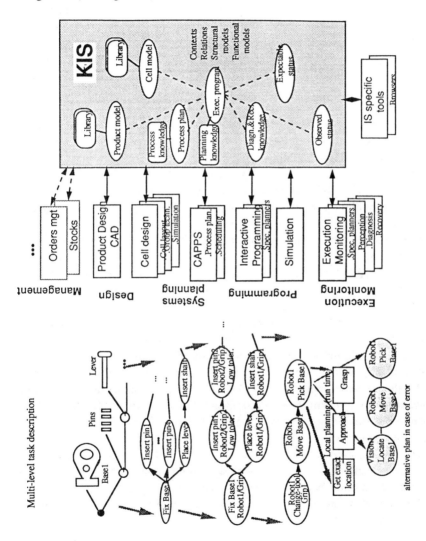

Fig. 21.1   KIS as the kernel for integration in a multi-layered architecture

21.1.2 Knowledge-centred approach

In order to support the implementation of decision-making processes (advanced cell controllers or decision support systems, for instance), integration should be knowledge-centred [2][5][8].

Knowledge Management Systems offer greater expressiveness than more classic technologies like DBMS, therefore getting closer to the representation requirements of engineering applications. In such domains, a capability to represent a number of complex concepts and objects, as well as a set of varied relationships among them, is required. In open systems, where the concept of integration and its implications has not been completely understood yet, flexibility in terms of the representation structures is highly desirable (fast prototyping). Existing knowledge management tools, combining various representation/ programming paradigms, are good candidates for the integration kernel.

On the other hand, to implement effective decision making subsystems, it is not enough to have 'good' (rich) KBs. In complex real systems, multiple knowledge-based decision making components can be identified. These components must be driven by real data. As mentioned before, such information originates in multiple sources supported by different technologies (CAD, CAPP, DBMS, cell controllers etc.). Therefore a KB system can act as the 'collector' for those information sources, providing a sort of uniform and centralized information 'front-end' (see Fig. 21.2) [2][15].

Additionally, one must note that this knowledge and information system has multiple 'users' (functional modules and human experts), requiring different views and abstraction levels. For instance, the set of characterizing attributes of a robot required by a cell planner is not the same as the model needed by a cell programmer.

Another motivation for a knowledge-centered approach comes from the need of translation mechanisms to generate/interpret various representations [3]. The information originating from the aforementioned sources does not come in the most structured way, or is based in different representation paradigms and restricted contexts, which demands a system with some degree of 'intelligence' and great flexibility in the modelling facilities. Consumers, on the other hand, require information in particular representation forms, implying a flexible translation process.

All these arguments justify the option for using a Knowledge Management System, presenting multiple representation/programming paradigms as the kernel for the information integration process in advance systems. Fig. 21.2 shows an information system development realized in accordance with this approach.

232

21.2 Development environment

21.2.1 System architecture

In a typical CIM environment, the information needed for programming and controlling a robotic cell comes from multiple sources and will be 'consumed' by multiple modules. These producers and consumers are supported by different technologies (CAD systems, DB management systems, cell controllers, CAPP systems, etc.), coming from multiple vendors and running on a distributed environment. Therefore, as a starting point for the development of the proposed architecture, a development platform giving support for information integration was established [8].

An integration of potential sources/consumers support subsystems was realized at UNL around a knowledge base management system (Knowledge Craft) offering multiple paradigms: frame representation, reactive, object-oriented and rule based programming.

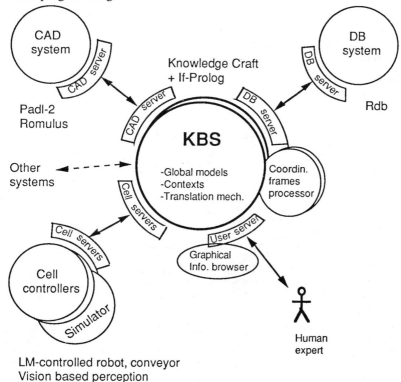

Fig. 21.2 KIS development environment.

233

In order to allow an adequate use of each component according to its specialization and simultaneously keeping a great accessibility to each information element, a tight coupling approach was chosen. Each connection implied the development of specialized servers responsible for the dialogue (under a defined protocol) and translation of information representations, e.g., to translate a table based representation (DB) into a frame based structure (KB).

An interactive approach for these connections was followed. That is, an initial loading into the knowledge base of information held by a subsystem can be partial. Additional elements can, interactively, be transferred on demand. The reactive programming paradigm proved to be adequate for the establishing of such interfaces. For instance, when an attempt to read the value of an attribute of a part (represented by a slot of a frame) is made, an if-read demon is implicitly fired and requests the necessary information from the CAD server. This mechanism provides a uniform and centralized interface to the information/services offered by heterogeneous subsystems, facilitating the development of a global control system. At the centralized system one does not have to bother with the peculiarities of each subsystem.

Figure 21.3 represents a distributed architecture integrating multiple information sources which was used to support one demonstrator system (High Level Interpreter), a joint effort of the Universities of Lisbon, Karlsruhe and Amsterdam [4].

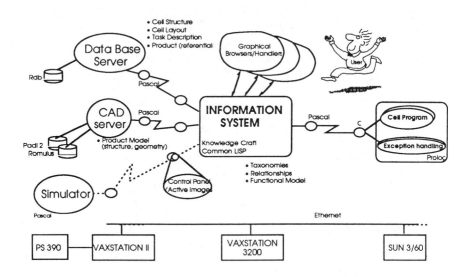

Fig. 21.3   Distributed Information Integration System.

21.2.2   Subsystems

- Connection KB – CAD

This connection enables the integration of geometric information (product data, for instance), provided by a CAD system, into the IS. Interfaces to two solid modellers – PADL-2 and ROMULUS – were developed [10] [6]. In establishing the communication protocol, some concepts were borrowed from CAD*I. However, one basic difference between CAD*I and our system is that we pursued an interactive connection while CAD*I intended to standardize information exchanges between CAD systems via neutral files.

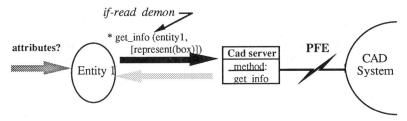

Fig. 21.4   Example of interactive connection KB - CAD.

In this realisation some structural information is loaded into the KB when the connection is established. Specific attributes for a given object may be retrieved on demand from the CAD system in a transparent way: when a request to such information is made on the KB side, a demon attached to the object is implicitly fired and queries the CAD system (Figure 21.4).

- Connection KB - DB

Relational data base management systems, in spite of various limitations to support engineering information, have achieved a growing level of utilization in industrial environments mainly as a consequence of being a "stabilized" technology. This is also the case with the industrial partners in our project consortium.

This kind of connection allows the access to such information. Current implementation uses Rdb [9].

When a DB is open, a structure of concept frames is built on the KB side according to a user specified mapping. Tuples selected from the DB are represented as instances of these concepts. These instances may, on the other hand, be real or virtual. In the first case slot values are loaded, on creation time, with data read from the DB. The virtual instances remain 'empty' and when any of

235

their slots is accessed, an indirect call – via reactive programming – to the DB is performed (Figure 21.5).

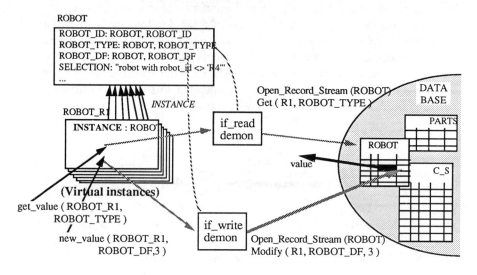

Fig. 21.5   Connection KB - DB.

- Connection KB – Prolog based modules

To extend the reasoning capabilities and to facilitate the communication with cell planning/programming and exception handling modules of the aforementioned High Level Interpreter demonstrator, an interface between KC and IF-Prolog was established.

Cell programming and exception handling modules, developed in IF-Prolog, run on a SUN 3/60 under UNIX. These modules interact with the IS via Ethernet using an interface written in C that exchanges messages with a Netserver implemented in Pascal on the VAX side. The communication primitives make available, on the Prolog environment, the CRL query functions of Knowledge Craft.

- Connection KB – cell components controllers

An interactive connection was also established between the knowledge based IS and a graphical simulator of robotic cells based on the PS 390 station (Evans and Sutherland) [3]. Each component is modeled as an object where its functional model is represented by a set of methods and demons. The associated Lisp functions perform the dialogue with the simulator.

A connection to real cell controllers (robot, vision based perception, conveyor) was also developed but, up to the moment, using a different KBS (frame engine developed in Prolog, running on a PC) (Figure 21.6). An implementation using Knowledge Craft was left to the next stage of development.

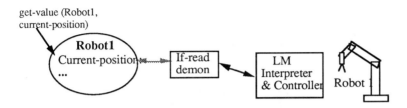

Fig. 21.6   Connection robot controller - frame based model.

The objective to develop a small frame engine kit in Prolog were twofold: Firstly, to get a system with the basic functionalities of a frame representation system but without the complexities of a general purpose tool (like commercial systems), and secondly to provide a simple way for the training of the group members on knowledge representation concepts before dealing with all complex peculiarities of heavier systems.

- Connection frames processor
Support to represent coordinate frames (referentials) and the relationships among them, as well as a set of transformation operators is an important aid for the High Level Interpreter. In this way, planning modules may manipulate referentials at the symbolic level, while the actual values are kept by the underlying processor. A flexible coordinate frames processor, based on the reactive programming paradigm, was developed in Knowledge Craft and integrated into the development platform.

- Browsers
Finally, a set of graphical browsers allowing the inspection of (navigation by) the information structures and their modification were also developed resorting to the graphical facilities of Knowledge Craft. See Figure 21.7 for an example.

21.3 Examples of information structures

21.3.1 Cell model

The structures representing the cell model are:
1. Taxonomy of components (hierarchy of frames representing prototype models).
2. Cell structure and layout – a specific cell is composed of a set of instances of the above taxonomy components. These instances are generated from information originally obtained from a set of tables in Rdb, via the KB-DB interface, and coming from a cell planner subsystem.
3. Functional model – list of elementary (executable) operations for each component and interactions between components (Fig. 21.8). The elementary operations set is represented at the prototype model and inherited. This information is introduced interactively into the IS via a graphical interface.

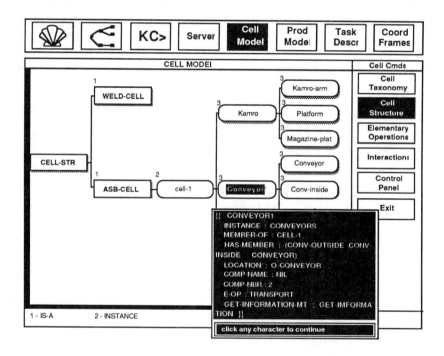

Fig. 21.7   Cell model.

238

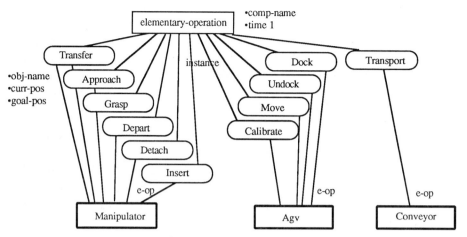

Fig. 21.8 Elementary operations.

21.3.2 Product model

Similarly to the cell model, there is a taxonomy of concepts to support the product description (structure, parts, and attributes). One specific product and its compon-

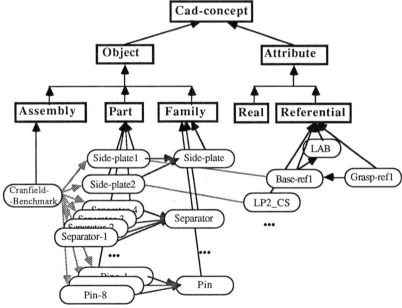

Fig. 21.9 Product model.

239

ents are represented by a set of instances of such concepts (21.9). This model is extracted from CAD (geometry data) and Rdb (relationships to cell components). All positioning information is represented using the coordinate frame processor.

The various models present in the IS are composed of information elements coming from multiple sources (Figure 21.10).

In the example, the product model is built up in the Knowledge Craft environment using the attributes provided by a CAD system, by a relational DB and by the user (via an interactive interface).

To facilitate the integration of multiple information sources, an "open system" philosophy has to be followed. One of the main questions at this level is the lack of standards for information interchange. For instance, for CAD information, some concepts of CAD*I were used, but this standard proposal is not sufficient for the required information. Emerging proposals like STEP may give a contribution to solve the problem.

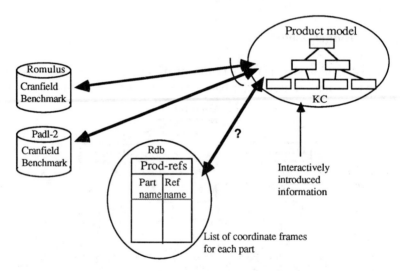

Fig. 21.10 Product model – an example of the addictive perspective for the information integration.

21.3.3   Task description

A task description at an implicit level is represented by a direct graph of abstract operators (precedence graph/process plan), Fig. 21.11.

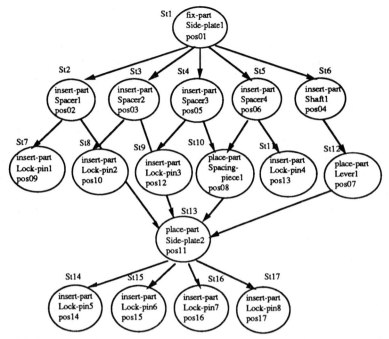

Fig. 21.11 Example of abstract task description.

This abstract plan is detailed by the programming modules until an equivalent description using elementary (cell executable) operators is reached (Fig. 21.12).

The task program is also represented as a directed graph, or a set of lists of operations, one list for each component (robot, conveyor, vision, etc.). Besides the operator type and corresponding parameters, each node of this program has slots to store the monitoring conditions and recovery strategies generated by the exception handling module.

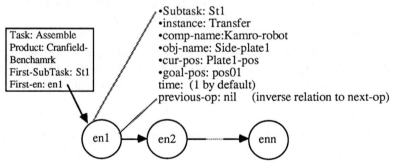

Fig. 21.12 Explicit task representation.

21.4   Conclusion

The initial stage of this integration work has led to a better understanding of various inter-related fields of CIM (design, systems planning, programming and execution monitoring) and a clearer identification of needs and requirements. A sound concept of components integration based on an Information System was the instrument for fruitful evolution [14].

However, the complexity of the problems imposed a development confined to a limited environment – only robotic cells were considered. Additionally, realized prototypes were somewhat tailored to specific demonstrators and therefore presented some weaknesses.Some of the main questions are related to the generality of the knowledge models, use of standards for modelling and information exchange, normalization of user interfaces (in different modules).

References

[1]   L.M. Camarinha-Matos, *Plan Generation in Robotics – state of the art and perspectives*, ROBOTICS (North-Holland) **Vol. 3**, No. 3&4, 1987.

[2]   L.M. Camarinha-Matos, *Sistema de programcao e controle de estacoes roboticas - uma aproximacao baseda em conhecimento*, PhD Thesis, Universidade Nova de Lisboa, 1989.

[3]   L.M. Camarinha-Matos, J. Moura-Pires, H. Pita, *Human oriented control of robotic cells*, ROBCON 5, Varna Bulgaria, Oct 1989.

[4]   L.M. Camarinha-Matos, U. Negretto, G.R. Meijeer, J. Moura-Pires, R. Rabelo, *Information integration for assembly cell programming and monitoring in CIM*, 21st ISATA - Int. Syp. on Automotive Technology and Automation, Wiesbaden, Germany, 6 - 10 Nov 1989.

[5]   L.M. Camarinha-Matos, A. Steiger-Garcao, *Robotic Cell Programming: A Knowledge-based Approach*, Proceedings of 8th IASTED/AFCET International Syposium on Robotics and Artifical Intelligence, Toulouse 18 - 20 Jun 1986.

[6]   L.M. Camarinha-Matos, A. Steiger-Garcao, *An Information System Architecture for Robot Cell Programming*, Nato Advanced Study Institute

onn CIM: Current Statuts and Challenges, Istanbul, Turkey Ago 31 - Set 12, 1987 (Springer-Verlag, Nato ASI Series, 1988).

[7]   L.M. Camarinha-Matos, A. Steiger-Garcao, *An Integrated Robot Cell Programming System*, International Syposium and Exposition on Robotics (ISER/19th ISIR), Sydney Australia, 6 - 10 Nov 1988.

[8]   L.M. Camarinha-Matos, A. Steiger-Garcao, *Knowledge architecture for flexible programming of robotic cell*, 20th ISIR, Tokyo, Japan, Oct 1989.

[9]   D. Ferreira, L. Osorio, *Interactive link between a database and a knowledge base: a prototype application in a CIM context*, UNIDO Workshop on Robotics and CIM, Lisboa, 11 - 15 Sep 1989.

[10] J. Moura-Pires, L. Osorio, *Knowledge based product model supported by an interactive link to CAD systems*, UNIDO Workshop on Robotics and CIM, Lisboa, 11 - 15 Sep 1989.

[11] A. Steiger-Garcao, L.M. Camarinha-Matos, *A Knowledge Based Approach for Multisensorial Integration*, Proceedings of NATO Advanced Research Workshop on Languages for Sensor-based Control in Robotics, Castelvecchio Pascoli, Italy, Sep 1 - 5, 1986. edited by Springer-Verlag, Nato ASI Series N. 29, 1987.

[12] A. Steiger-Garcao, L.M. Camarinha-Matos, *A Conceptual Structure for a Robot Station Programming System,* ROBOTICS - International Journal (North-Holland) **Vol. 3**, No. 2, Jun 1987.

[13] A. Steiger-Garcao, L.M. Camarinha-Matos, *An Integrated Architecture for Robot Cell Programming and Monitoring*, 2nd International Symposium on Robotics and Manufacturing Research, Education and Applications, Albuquerque, USA, 16 -18 Nov 1988.

[14] A. Steiger-Garcao, L.M. Camarinha-Matos, *Research and industrial exploitation perspectives on Robotics and CIM*, UNIDO Workshop on Robotics and CIM, Lisboa, 11 - 15 Sep 1989.

[15] A. Steiger-Garcao, L.M. Camarinha-Matos, *Information and knowledge integration for CIM – concept summary*, 21st ISATA - Int. Syp. on

Automotive Technology and Automation, Wiesbaden, Germany, 6 - 10 Nov
1989.

[16] A. Steiger-Garcao, L.M. Camarinha-Matos, *Knowledge-supported
autonomy for robotic systems in CIM*, IAS2 - Intelligent Autonomous
Systems 2, Amsterdam 11 – 14 Dec 1989.

Chapter 22

Conclusions

A. Steiger-Garcao, L.M. Camarinha-Matos
Universidade Nova de Lisboa

The integration of robotic systems in a CIM environment requires the integration of heterogeneous tools/modules for cell design, planning, simulation and supervision. The information-based approach for integration has proved to be an adequate one, as shown by implemented demonstrator prototypes. The use of an Information System as the kernel for integration has been widely accepted in designing CIM systems in general. This project showed the approach to be also adequate at the robotic subsystems level as well as for the integration of such systems into a more general framework.

In terms of realized prototypes, two implementation approaches were explored:

i) Data base approach with extensions towards the management of the overall design or planning process (Chapter 20).
ii) Hybrid knowledge-base approach combining different information management technologies - KBMS, RDBMS, CAD, etc. (Chapter 21).

In spite of these apparently diverse directions, common models were pursued at the conceptual level. That is the case for cell/components models (functional, structural, kinematic, dynamic models), task description, program representation, product model, etc. The exploration of different implementation approaches for common conceptual models allowed a better clarification of advantages and limitations of available information management technologies concerning engineering needs. Of particular concern in the described work were the aspects of:

- Modelling – expressiveness capabilities to represent a number of objects and complex concepts as well as a set of varied relationships among them;
- Platform for decision support systems and interactive decision making;
- Fast prototyping – strongly needed in a context where the concept of integration and its implications has not been fully understood yet;
- Translation mechanisms to generate/interpret various representations and views.

As in other application fields of Information Technologies, acquired experience in the robotic systems domain points to the need of an evolutionary convergence of both approaches towards an information management system showing powerful representation mechanisms, such as:

- Concept of object, hierarchies and other structural relationships;
- Flexible inheritance;
- Support for behavior modelling (methods and demons);
- Flexible constraints specification and rule-based inference;
- Graphical browsing and tutoring/intelligent explanation facilities;

as well as some more classic functionalities:

- Integration of different views;
- Concurrency and consistency maintenance;
- Long-term storage and efficient information retrieval;
- Standardized query language;

On the other hand, the implementation of different levels of tools' interconnection – ranging from loosely coupled to tightly coupled systems – pointed to the strong need for a standardized integrating infrastructure. Existing technology still raises great difficulties when attempting to integrate distributed heterogeneous systems. Other works carried out by some ESPRIT projects – integrating infrastructure of CIM-OSA, CNMA, ISA – are likely to improve this situation.

In summary, the main challenges for a next phase in CIM and robotic system integration include:

- Establishment of standards for information exchange. STEP activities are expected to play an important role in this field.
- Establishment of integrating infrastructures for distributed heterogeneous systems, including standardized access methods in the Information System.

- Development of better tools for information modelling, and information representation and management.
- Development of a general taxonomy of CIM Engineering Activities and a Common Glossary, as a first step towards a wider modelling normalization.
- Evolution towards CIM generative systems, i.e., flexible toolboxes, driven by a meta-knowledge base and able to generate specific information systems for particular architectures (CIM CASE).
- Design and specification of flexible administrative policies for software configurations (flow of tools' activation, concurrency constraints, version-keeping strategy, etc.), tailored to specific installations.
- Development of more effective knowledge and information acquisition tools, possibly resorting to recent results coming out of the Machine Learning community.

Section V

Applications of the system

Chapter 23

Introduction

R. Dillmann
University of Karlsruhe, Germany

In the previous sections of this book the basic system components supporting systems planning, programming and information management have been presented and discussed. To prove and to evaluate the efficiency of the integrated approach for planning and programming of manufacturing tasks, different applications ranging from palletizing, welding, laser cutting and assemblies have been implemented. They are described in detail in the following chapters. Two approaches have been followed. One assembly application has been realized to show the challenge of further implicit task-oriented planning and programming systems. The Cranfield Assembly Benchmark has been consequently modelled, planned and automatically mapped onto a robot manufacturing cell (Chapter 25). The other applications have been developed using interactive approaches for planning and programming to realize a matrix printer assembly, sheet metal welding, laser cutting (Chapter 26), space robotics (Chapter 27 and 29), pallerizing of clutch cases (Chapter 28), underwater robotics (Chapter 29). These applications are described by the partners Renault, FIAR, KUKA and IPK. It can be shown that various specific application-dependent technological aspects could be integrated with the help of an information management system. In both cases, planning and programming of the manufacturing tasks involves the representation, the specification and the modelling of the product, of the system components and of the manufacturing cell operation. For this purpose, a chain of tool-supported activities producing and consuming information has been applied. Process planning for manufacturing requires that a product's design must be translated into the best representation for the product's manufacture. The manufacturing strategies considered range from full hard automation, flexible line manufacturing, flexible cell structure and manufacturing-insula-oriented manufacturing. Planning, programming and robot process control have to consider the selected strategy. Manufacturing process plans consist of information about the manufacturing process to be applied, its process parameters, the machines to be used and the time schedule. Synthesis of the process information and the use of decision logic has been performed by a human expert interactively supported by the software tools of the consortium or by highly sophisticated automatic implicit task-oriented software systems. Integrating CAD, CAM and Robotics requires the use of database management systems which support all activities of planning, programming and process control with appropriate information. A standard representation scheme for the different information and object classes used for planning and programming facilitates the common sharing of information. With the help of relational data base management systems, consistent object classes have been modelled. Among the partners various cell models, product data and task information have been exchanged. The activity chain from product design via planning activities to the product manufacturing plan information and the specification of elementary robot cell component actions are shown in the following application descriptions. The individual steps for constructing an executable robot cell program are illustrated. The use of graphical simulation systems and data base management systems are illustrated in each application.

Standards under work in key projects of ESPRIT CIM have been considered and applied if available. For ESPRIT 623, the European Computer-Integrated Manufacturing Architecture (AMICE, Project No. 688), the Communication Network for Manufacturing Applications (CNMA, Project No. 955) and CAD Interfaces (CAD-I, Project No. 322) have been considered. Other projects related to Open CAM Systems (Project No. 418), Control Systems for Integrated Manufacturing (Project No. 477), Knowledge-Based Real Time Supervision in CIM (Project No. 932), General Purpose Sensory-Controlled Systems for Parts Production (Project No. 118), and many others influenced the implementation.

Areas of standards useful for integration and partly implemented by the partners are related to:

- Communication;
- Networking;
- Product description;
- Manufacturing process control;
- CAD interfaces;
- Graphics;
- Data base management systems;
- Programming languages for operational control; and
- Standard tool box systems.

The different manufacturing tasks implemented all implied the determination of basic manufacturing operations, the technology to be applied, the process parameters, and the constraints to be considered. Sequencing of the elementary operations as well as the specification of parallel operation and their sychronization has to be planned. The resulting plans can be logically tested and optimized in terms of manufacturing time cycles. A number of planning systems have been developed by the project partners. They fall into two basic categories. Some planning systems are based on retrieval of standard plans of coded products applying a product variant system with parameters and classifications among the main elements. This variant approach requires that the existing process plans must be summarized and coded to be classified under similar part families. The other category of planners follow a generative approach which allows the creation of a new plan for a new robot manufacture. For robot assembly operations, assembly graphs and precedence graphs are deduced from the product CAD information. From these graphs the basic assembly operations and the time sequence of their execution can be derived. The basic assembly operations can be further refined and broken down into elementary robot operations like grasp, transfer, insert, etc., which can further be used for robot cell programming.

Programming of robot-based manufacturing cells is characterized by a wide spectrum of programming techniques. In this project the unique ESR representation and the ESL language have been used, which allows, with the help of a code generator, the generation of different robot codes. The purpose of these is to specify sequences of robot movements and endeffector interactions, which may be sensor-guided. The resulting movement can be animated for the user on a graphical screen. Simulation techniques facilitate the test of the robot program. The simulator is referencing the world model and the program information. All partners have installed graphical simulation systems which are partly CAD-system-dependent or CAD-system-neutral.

Modelling, planning and programming can be performed more efficiently and can be further automated if the produced and consumed data are consistent and can be retrieved. Thus, a key role for system integration is dedicated to the central database management system as shown in the following application descriptions. It supports all modules needed for robot programming with the required information.

The activities outlined above are described in detail in the following chapters where different applications are planned and programmed following an integrated operational control system approach. Finally Chapter 30 presents the conclusion to Section V.

Chapter 24

Planning and programming of a dot matrix printer assembly

G. Stark
KUKA Schweißanlagen & Roboter GmbH, Germany

R. Bernhardt, V. Gleue, S. Krüger
IPK Berlin, Germany

24.1 Introduction

The main objective for realizing demonstrator systems by integrating modules of different partners was to show the increased functionality and efficiency of an integrated planning and programming procedure for robotized manufacturing cells.

Thereby the different principles such as automatic or interactive planning functions and explicit or implicit programming procedures were considered. Additional realistic industrial applications or well-known benchmark tests were selected. A further important criterion for the selection of applications was to cover a broad spectrum of problems and to show their solutions. Especially for the integrated system described in this chapter, an assembly application was selected considering

- parts of very different size requiring gripper exchanges,
- parts of different shapes requiring multifunctional grippers, and
- assembly operations with different mating directions requiring complex fixtures and robots with six degrees of freedom.

The realized systems or – depending on the production task – parts of them are being used by the project partners for a variety of industrial applications.

The principle tasks of the integrated system are:

- detailed planning of an assembly process;
- selection of all components and their arrangement;
- generation of robot application programs and
- simulation and test of the assembly task execution within the planned work cell [1, 2].

The integration of components and tools was achieved primarily through information exchange and management, performed via a relational database system. This enabled a safe data handling and led to a most flexible system structure with regard to the integration of functional system units and a quick adaption of user needs.

In the following, an example of an application of some of the planning and programming tools developed in the context of this project is given. The production task that has to be executed is a subassembly of a dot matrix printer. The parts to be assembled are the following:

- the plate of the printer chassis;
- the stepper motor;
- the gear;
- the rocker;
- the lock washer;
- two screws.

In Figure 24.1 the assembled matrix printer line feeder is shown.

To assemble this drive unit of the printer's line feeder, two interlocking drive gears must be positioned. Before the stepper motor is screwed to the plate of the chassis, the gears of the motor pinion and the rocker must be positioned and interlocked. This requires moving the motor along an oblong hole in the chassis' plate, limiting the motor's range of motion. After the motor gear and rocker have been joined correctly, the motor is screwed to the plate. This application example has been selected for various reasons:

- It is an actual production task that is being planned by one of the partners in the project.
- The example is simple enough to be taken as an initial application, but complex enough to produce meaningful feedback.

-   Different grippers and parts with varying shapes and sizes are involved.

The following sections are to demonstrate how the tools, developed in the context of this project, are used in order to plan this application.

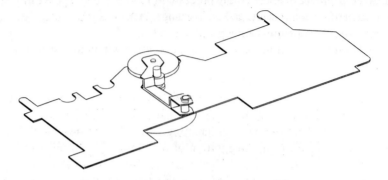

Fig. 24.1   Assembled matrix printer line feeder.

24.2   Assembly sequence planning

The integrated procedure starts with the generation of an assembly sequence plan which includes the operation sequence planning, the precedence analysis, the preselection of robots which are suitable for the different assembly steps and the selection of available feeders and grippers. Figure 24.2 shows in decomposition the parts whose assembly sequence has to be determined.

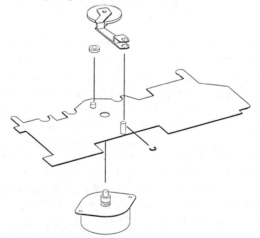

Fig. 24.2   Parts for assembly sequence planning.

256

24.3 Component selection and layout arrangement

Based on the assembly sequence and the suggested device, the components are selected and roughly arranged. During an interactive and iterative procedure this layout is improved and verified using support tools for, e.g., cycle time estimation and material flow simulation. Figure 24.3 depicts the selection and first arrangement of the components within the layout.

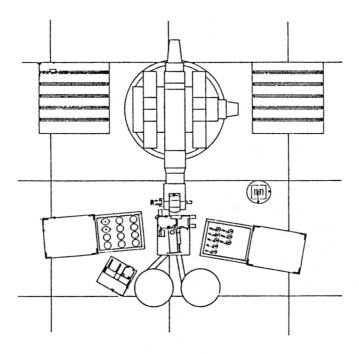

Fig. 24.3   First layout arrangement.

24.4   Assembly process planning

In the next step, the generated planning data are transformed into a structure suitable for the offline planning procedure. Besides the layout description, the assembly sequence is described with a semiformal representation (similar to PSL), containing also geometrical, technological and control-specific data. Fig. 24.4 shows the generated assembly process plan.

Nr.	Object	Action	Task	Event	Comment
48	rocker	speed 20 move P23 to P24			mating point
49		deactivate gripper BII			mating rocker
50		move P24 to P23			depart
51		speed 100 move P23 to P25		start rotate gripper BIII	
52		speed 20 move P25 to P26			picking lock washer
53	lock washer	move P26 to P25			
54	lock washer	speed 100 move P25 to P27	orientate gripper B		
55	lock washer				mating lock washer
56		move P28 to P27	turn back gripper BIII		depart
57		speed 100 move P27 to P16	orientate gripper B		

Fig. 24.4   Assembly process plan.

24.5   Optimization

The planning phase is completed with a layout optimization procedure considering type variants and locations of robots, tools, and peripheral devices and their influence on important parameters, reaching from cycle times up to the total cost of the work cell. The result is the final definition of the layout, shown in Figure 24.5.

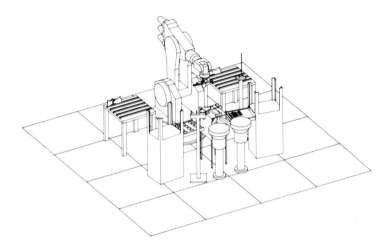

Fig. 24.5   Final definition of the layout

24.6   Robot Execution Planning

Based on the general description of the assembly task and the optimized layout, the off-line programming procedure for the robot is performed. As a first step the execution of the task by the robot has to be planned. This is related to the determination of the motion types and parameters, the conversion and integration for technology information, as well as to the scheduling of the task and the explicit solution representation (ESR) which is a data structure containing all necessary information to generate robot application programs in a specific language. Figure 24.6 presents the track in space that the robot generates when moving the plate.

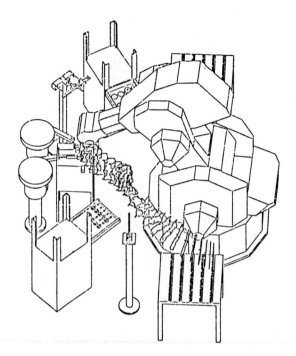

Fig. 24.6   Robot moving the plate.

24.7   Robot program generation

In the next step, out of the ESR a robot program will be generated automatically. For the first realization, a subset of the SIEMENS robot language SRCL has been used. The executability of the off-line created application programs has to be tested, i.e., to verify whether or not the assembly task is fulfilled. Figure 24.7 shows the generated program for the target system.

24.8   Process execution simulation

Due to the complexity of the necessary decision-making process required for verification, the human operator has to be provided with a presentation of the task execution as realistic as possible. This is achieved via the simulation system by visualizing the robot motion as well as the actions of peripheral devices of the assembly cell. Additionally robot controller-specific cycle times and joint parameters are measured.

```
┌──────────────────┐
│ Robot Task       │
│ Representation   │
│                  PRGO
│                  PAGO START
│                  OBFIX GZ
│                  PAGO GM
│                  OBUNFIX GZ
│                  ...
│                  ...
│                  ...
│                  PRFINISH
└──────────────────────────────┘

            ⇓

┌──────────────────────────────────────────────────┐
│ SRCL  Program                                     │
│        \*..........................................*
│        \*    File :   ROB.src                     *
│        \*    erzeugt am  29-SEP-88  aus  KUKA_TEST3.SRC  *
│        \*..........................................*
│        DEF HP50
│        ORI VAR
│        BES ALL 100
│        BES BAN 100
│        GES ALL 100
│        TXT GREIFER AUFNEHMEN
│        WIS32
│        UES BAN 0
│        PTP X+1330.0 Y-507.7 Z+1211.5 A+150.000 B+0.000 C+180.000
│        GES BAN 101.250
│        WIS32
│        UES BAN 0
│        LIN X+1330.0 Y-507.7 Z+1191.5 A+150.000 B+0.000 C+180.000
│        TXT GREIFER ANKETTEN
│        S A 26
│        GES BAN 101.250
│        WIS32
│        UES BAN 0
│        LIN X+1202.7 Y-434.2 Z+1191.5 A+150.000 B+0.000 C+180.000
│        ...
│        ...
│        ...
│        END HP50
└──────────────────────────────────────────────────┘
```

Fig. 24.7   Generated program for the target system.

24.9   Conclusion

This chapter has described the application of planning and programming tools developed within ESPRIT 623 to the assembly of part of a dot matrix printer. Each of the steps described are the subject of previous chapters of the book as follows: assembly sequence planning (Chapter 6), component selection and layout planning (Chapter 7), process planning (Chapter 8), layout optimization (Chapter 9), robot execution planning (Chapter 12), robot program generation (Chapter 13), and process execution simulation (Chapter 14). Thus these steps represent a progressive set of activities in an planning and programming system for robot based assembly.

261

References

[1] R. Bernhardt, *Integrated Planning and Off-Line Programming System for Robotized Work Cells*, Proceedings of the 6th Annual ESPRIT Conference, Brussels, 1989.

[2] G. Stark, *Operational Control for Robot Integration Into CIM and its Applications*, IFAC Conference 'SYmposium on RObot COntrol' (SYRO-CO), Karlsruhe, 1988.

Chapter 25

Programming of the Cranfield assembly benchmark

K. Hörmann, U. Negretto
University of Karlsruhe, Germany

The Cranfield Assembly Benchmark was chosen as the assembly task to demonstrate the functionality of two developed systems described below. The systems address two different application fields in the framework of robot-integrated assembly environments:

- the automatic off-line program generation for one or several cooperating robot arms and
- the control and scheduling of a robotized cell with exception handling.

25.1   Task-level programming

The task-level programming system is an off-line programming system in which robot actions are specified only by their effects on objects. The resulting sequence of robot motions are generated automatically by the system. One of the realized applications of the system is the so-called Cranfield Assembly Benchmark which consists of 17 parts which can be assembled to form a mechanical pendulum (see Fig. 25.1). The task-planning process is performed in the following steps :

- At a workstation, supported by a comfortable graphic interface, a user interactively describes a workcell. Normally, the workcell components are standardized parts and therefore may be selected from a predefined library.
- Then, the user graphically specifies the assembly task to be performed and starts the planning system.

The system automatically generates a sequence of safe robot actions/motions necessary to perform the task. It translates these robot instructions into executable robot commands specific for the robot type and passes them to the simulation system which visualizes the robot motions found for the specified assembly task. Therefore, the actual presence of the real robot or of the workcell is not needed.

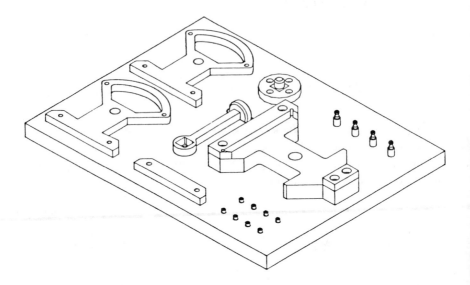

Fig. 25.1   The Cranfield benchmark for the assembly of a mechanical pendulum in its initial position.

The basic interactions between the component modules constituting the system are shown in the reference model (see Fig. 25.2). The user specifies the goal positions of all the work pieces and the contacts between them (lay_on, inserted, completely inserted) supported by the graphical editor. The graphical editor requires a FEMGEN – description of the objects in the workspace of the particular robot. This description is used to construct the internal object description based on a boundary representation form. The goal positions of the parts as well as the types of contacts between the parts in goal position are stored in a disk file.

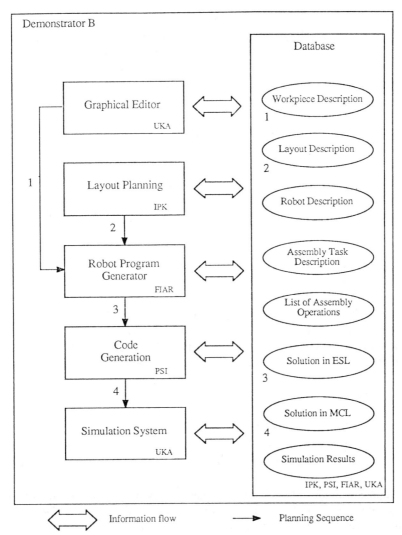

Fig. 25.2   Reference model.

The Robot Program Generation System is a tool for automatic off-line robot program generation for assembly tasks. It generates programs for several robots of a work cell which resolve problems of coordination and collision avoidance between the robots. Inputs to the system are a geometrical description of the parts to be assembled and the way they have to be connected (Task Description) and the definition of the workcell that has to execute the assembly task (World Model).

The system is composed of different modules whose functionalities are briefly described in the following:

- Assembly Graph Generator: This module determines, for a given assembly task, the order in which the various parts have to be assembled. It requires a geometrical description of the parts and the way they have to be connected. The resulting order is expressed by an 'augmented assembly graph'. Conventional assembly graphs do not state explicitly what operations can be executed in parallel or what sub-assemblies can be built for solving the assembly task. This module produces a precedence graph in which parallelism of operations is expressed by sub-assemblies and, in the explicit definition, which operations can be principally executed in parallel. This information is used to plan and coordinate the actions of the robots of the workcell to reduce the time needed for executing the overall assembly task.

- Scheduler: The Scheduler assigns assembly operations to the robots of the workcell using the ordering and the information about possible parallelism of operations determined by the Assembly Graph Generator. The goal is to produce a plan of operations for each robot of the workcell that is 'executable'. This means that it is actually executable by the robot and that it avoids collisions between the robot and the parts, or between two robots. Among the possible executable plans the Scheduler tries to determine the one corresponding to the minimum overall assembly time. The Scheduler estimates the time needed for assembly operations to synchronize the actions of the various robots.

- Program Generation and Motion Planning: Each operation of a plan determined by the Scheduler is expanded into a set of actions (for instance, 'assembly-insert' becomes 'approach, grasp, verify, transfer, insert, verify, release, depart'). Each action is expanded into a set of other actions, until the level of actual executable robot instructions (such as MOVE) is reached. In this process parameters of robot instructions are filled in such as positions to be reached, or trajectories to be followed.

- Code Generation: This module generates an ESR (explicit solution representation) which is a data structure containing all necessary information to generate robot application programs in a specific language. The code generator translates the ESR program into MCL (motion command language) which is understood by the simulation system ROSI.

- Simulation System: The simulation system ROSI is a software tool designed for the planning of robot applications by manufacturing engineers. The main hardware component is a high-performance graphic screen which achieves a comfortable man-machine interface. Using the graphic features of this interface, the engineer develops solutions of planning tasks for robot applications. The analysis of the off-line-developed task solution is supported by the graphic representation of the robot scene including a real-time animation of motions.

The simulation system ROSI is a suitable tool for the following off-line planning tasks:

- planning of the layout of manufacturing cells;
- selection of robots and end-effectors;
- generation of robot programs;
- graphical simulation of the program execution;
- visual detection of collisions.

ROSI is based on a modular concept. The modelling module is comprised of functions for modelling and defines geometry, kinematics and technology as well as functional parameters for manipulators and workpieces. For geometric modelling, the CAD-systems ROMULUS and EUCLID are used.

The planning of the assembly of the Cranfield Benchmark was realized for the following scenario. The assembly set consists of :

2 side-plates
1 lever
1 shaft
1 spacing-piece
4 spacers
8 locking-pins.

The assembly was performed by two Puma 260-robots working in parallel. The work-pieces were arranged on two base-plates lying on two tables. Each table was assigned to one robot and each robot only assembled the parts on 'its' table. The assembly was done on a third base-plate lying on a third table (see Figures 25.3 - 25.5). This layout was defined with the help of the modelling module of the simulation system. The resulting scene was loaded into the Graphical Editor. Then the user defined the goal positions and goal contacts of all the work-pieces using the features of the Graphical Editor. The resulting information was stored on disk files and the 'Automatic Robot Program Generation System' was started.

Fig. 25.3 shows a situation in which the assembly graph generator of this system determines the assembly direction for a locking pin. A step of the assembly graph found for the Cranfield Benchmark is given in Fig. 25.4. The Automatic Robot Program Generation System produced an ESR file as output. The code generator translated the ESR program into a MCL program which was executed and simulated by the ROSI system. An intermediate step during the assembly of a locking pin is given in Fig. 25.5.

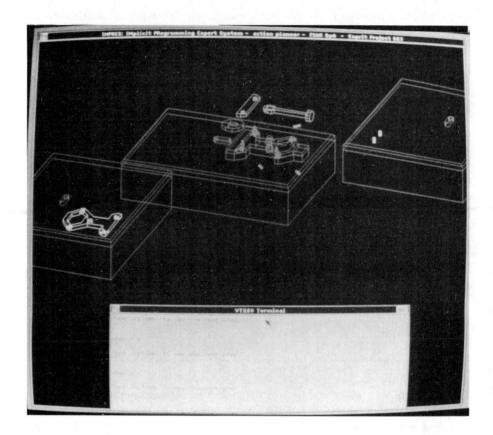

Fig. 25.3   Automatic determination and visualization of the assembly direction.

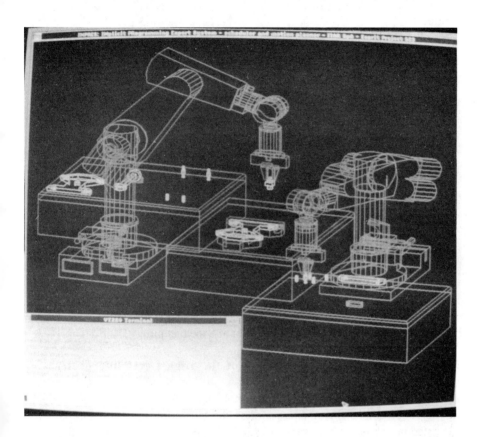

Fig. 25.4   Visualization of one assembly step defined by the assembly graph
which has been automatically generated.

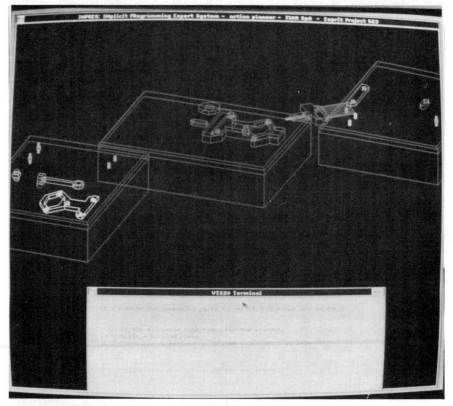

Fig. 25.5   Assembly execution simulation of a locking pin.

25.2   Planning and programming of the High Level Interpreter

The High Level Interpreter (HLI) is the implementation of a cell controller which, on the robot control level, incorporates the exception handling mechanisms presented in section 17.3. In section 17.6 the basic components of the HLI and the LLI are described. In this chapter we show how these components can be programmed. The realized prototype system also shows the integration of the HLI modules in the information system described in Chapter 21.

The objectives of the integrated system can be summarized as follows:
− specification of multi-robot production cell control;
− distributed information integration;
− task scheduling;
− monitoring and exception diagnosis;
− recovery planning;

– simulation, test and revision of the planned programs and exception handling strategies.

Previous chapters of section III, Chapter 24 and 25, have described how the planning and programming of robot motions can be realized with an off-line programming system. In particular, the geometrical aspects like path interpolation and optimization are taken into account in these systems. These planning and programming systems mainly address the functions of the LLI. What has not been handled is the control specification for flexible robotized cells, the task scheduling for the cell components and the planning of the exception handling functions of the HLI. The monitoring and diagnostics functions need to know which sensor systems are available and when to use which sensor for monitoring or diagnostics. The recovery planning functions need a description of applicable recovery strategies. This section will thus concentrate on the generation of the control model, the task scheduling, and the planning of the monitoring, diagnostics and recovery functions of the HLI.

The flexible assembly cell chosen as a testbed is shown in Fig. 25.6. Its components are the KAMRO (Karlsruhe autonomous Mobile Robot), a stationary Puma 600, a conveyor, a magazine and an assembly station. The sample task was the Cranfield assembly benchmark shown in Fig. 25.1.

The modules of the HLI in relation to the information system are shown in Fig 25.7. The first task is to build up the information structures and to fill the structure with the needed data. Out of the cell data the control model is generated. With the assembly task description given as precedence graph the cell program is generated and validated through execution simulation. The validated program is then expanded with monitoring conditions and exception handling strategies.

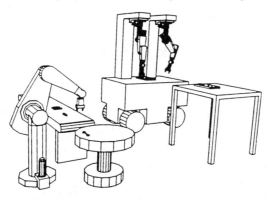

Fig. 25.6 The flexible assembly system.

271

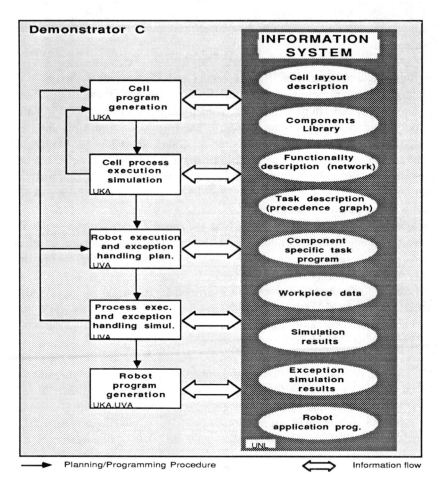

Fig 25.7   Functional reference model of the prototype system.

25.2.1   Information system

In a typical CIM environment, the information needed for programming and controlling a robotic cell comes from multiple sources and will be 'consumed' by multiple modules. These producers and consumers are supported by different technologies (CAD systems, DB managers, cell controllers, CAPP systems, etc.), coming from multiple vendors and running on a distributed environment. Therefore, the architecture in Fig. 25.8 was established to support the information integration.

272

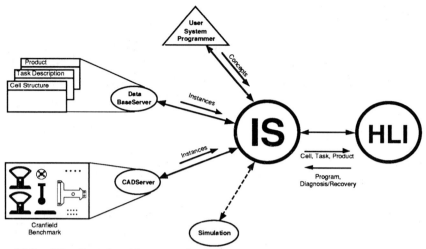

Fig. 25.8: Distributed architecture of prototype system.

The main components are:

- Connection KB – CAD;
- Connection KB – DB;
- Connection KB – Prolog-based modules;
- Connection KB – cell components controllers;
- Coordinate frames processor.

Information structures were developed for the information system to represent the primary input to the cell program generation module:

Cell model

The structures representing the cell model are:

- Taxonomy of components (hierarchy of frames representing prototype models).
- Cell structure and layout – a specific cell is composed of a set of instances of the above taxonomy components.
- Functional model – a list of elementary (executable) operations for each component and interactions between components.

273

Product model

Similarly to the cell model, there is a taxonomy of concepts to support the product description (structure, parts and attributes). One specific product and its components are represented by a set of instances of such concepts. This model is extracted from CAD (geometry data) and Rdb (relationships to cell components).

Task description

A task description at an implicit level is represented by a directed graph of abstract operators (precedence graph / process plan) (see Fig.25.9). This abstract plan is detailed by the programming modules until an equivalent description using elementary (cell-executable) operators is reached – i.e. the task program.

The task program is also represented as a directed graph, or a set of lists of operations, one list for each active component (robot, conveyor, vision, etc.).

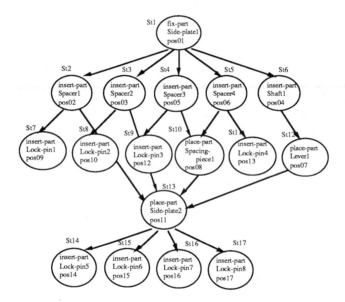

Fig. 25.9 Task description.

25.2.2 Cell programming and simulation

Multi-robot cell control systems are often dedicated to one specific task and are implemented on a very low level of detail, thus they are very inflexible. The reason lies in the fact that commonly used methods do not allow the specification of a task independent cell control system. To model the assembly cell environment in a task independent form requires the representation of the overall functionality of the cell. This functional representation has to include both static and dynamic properties of the cell. The static properties reflect the layout, the description of the components and the relations between the components, whereas the dynamic properties represent the synchronization and the concurrency of the processes. In the synchronization description the monitoring conditions have to be defined also.

The execution of the cell task then requires the representation of the task work plan on the functional model. This step is regarded as cell program definition. With the cell program and the functional cell model, the cell controller is able to monitor and to execute the process. The cell control system and the interfaces to the user, to the information system and to the graphic simulation system are outlined in Fig. 25.10 .

The programming environment to specify the cell controller, the program and the simulation of the process is separated into two modules:

– cell program generation;
– cell process execution simulation.

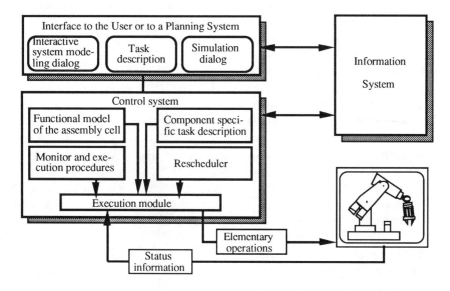

Fig. 25.10 Cell controller (cell level HLI) and interfaces.

The objective of the cell program generation module is to generate a cell model representing the functionality of the cell, then, based on this model, to generate a workplan for the task to be executed in the cell. The model constitutes the main part of the cell controller. Within the module, the two distinct tasks are worked out separately. Based on a pre-defined cell layout and on known device data, first the functional cell model is generated. Secondly, the explicit programs for the components of the cell are generated.

1. The definition of the functional model is performed by the user in an interactive system modelling dialogue. The key to a complete and exact representation of the system lies in the definition of the possible interactions between the single components. As mentioned, these so-called relations only depend on the functionality of each cell component (i.e., which operations it is able to perform) and not on the task to be performed. The relations strongly depend on the layout. Thus the layout description and also the component description with geometric data, kinematic data and technological data has to be known for the definition. This information retrieval is performed in a step prior to the modelling dialogue in a query session between the user and the Information System. After the information retrieval the user passes to the definition phase to build up the model.

   The model is based on a predicate transition net. Predicate transition nets with various extensions have already been used to model the static and dynamic properties of discrete systems, such as distributed systems, distributed data bases, flexible manufacturing systems, communication protocols, and others. Using this formal method, the functionality of each component is modelled in a predicate transition net, forming a local subnet. The components (i.e. subnets) are related to each other through synchronization transitions. The connection of all subnets is the overall net model. The overall net model represents the functional model of the cell describing the allowed concurrency, the synchronization between components, the material flow and the relations between the devices. The implementation of the description is realized in Prolog. The functional model can be changed interactively by declaring the relations in a Prolog formalism. The generated model is stored in the information system and serves as input to the following module.

2. The generation of the cell program is performed automatically. The planning module generating the program takes as input the functional model of the cell and the description of the assembly task.

An assembly consists of a number of sub-tasks (abstract operators) which have a dependency relation given by the application. The dependency relations indicate the necessary order in which two tasks have to be carried out. The order constraints can be of a geometrical nature like the assembly problem of stacking several parts upon each other. Also, the availability of parts or production resources can impose constraints on the allowable order of the tasks, see Fig. 25.9.

The function of the cell process execution simulation module is to realize a simulation system for cell program execution. The simulation system offers tools for testing and validating the planned cell program. The system is based on a time-driven interpreter and is currently tested on an operation driven mode. It enables the simulation of the cell program flow and offers various possibilities of analysis. The results improve the quality of the former generated program in respect of performance time. The system also detects deadlocks due to unsolvable conflict situations or because of material flow reasons. The user is able to generate interrupts in the execution simulation to verify the system state or to change a set of parameters and check the effect of the change after restart of the simulation.

The simulation module presents the status of each component with the actual active operation and a preview of the next operations to be performed. The trace flow of the execution is presented to the user and is updated at each start of a new operation of one or more (i.e. parallel) components.

25.2.3   Execution monitoring and exception handling

This section will concentrate on the planning of the monitoring, diagnostics and recovery planning functions of the robot level HLI.

The programming environment is based on the model presented in Chapter 17.3.4. It consists of two programming phases and a post processing step to realize the coupling of the physical robot to the control system. The planning steps are:

I   -   Exception handling planning by expansion of robot program.
II   -   Exception handling simulation and tailoring to application.
III -   Generation of robot specific instructions.

I) In the exception handling planning phase, the elementary operations of the robot program are automatically expanded with instructions for the monitor, fault trees for the diagnoser, and handling strategies for the recovery planner. The underlying idea is that for a specific application area and robot-sensor configuration the required functionality of the exception handling functions is similar. For example, if the robot and available sensor systems are known, the exception handling functions needed for supervision of an assembly system will not differ too much between two such assembly tasks.

By taking the elementary operations as the smallest unit of operation, a preselected set of instructions for the monitor, diagnoser and recovery planner can be made. This set forms the kernel which is automatically added to the representation of each elementary operation of the robot program. The specification of the application-independent kernel is carried out by a system programmer. Based on expert knowledge of a particular application field like assembly or welding, a set of strategies for possible exceptions is specified. The system programmer identifies the possible exceptions which can occur during the execution of an elementary operation and specifies the handling strategies for all these cases.

The first step in expanding the robot program planned by the cell program generation module is to add the information needed by the monitor. As described in 17.6, for each elementary operation a list of monitoring conditions is specified. This list instructs the monitoring module on which sensor primitive should be checked when the corresponding elementary operation is executed. In the planning phase the required value of the sensor primitive is derived from the parameters of the corresponding elementary operation. For instance the required status of the gripper is determined by using the value of the 'Part' parameter of the elementary operation. If this parameter is nil, no part is expected in the gripper. If the parameter contains the name of a part, the width of the gripper is determined from the geometrical description of the part. The sensor primitives are listed in Table 17.6.2.

Elementary operation	Sensor primitive		
	pre	during	post
Transfer	OA, OO	CM, RF, AO, OO	PP, AO, OO
Grasp	RF		PP, AO, OO
Detach	OA, OO		PP, RF
Insert	RF, AO,OO	RF, OA, OO	PP
Approach	AO, OO	RF, OA, OO	PP
Depart	RF, AO, OO	RF, OA, OO	PP

Table 25.2.1 Elementary operations and monitor conditions.

The codes CM, RF, PP, AO and OO represent sensor primitives and are described in Table 17.6.2. In Table 25.1 the elementary operations of the robot and the corresponding monitor conditions are given. The codes refer to a check of a sensor primitive. For instance 'OA' should be read as:

'Check value of Object Available Primitive ≠ Expected Value'.

The monitor uses this table to select the sensor primitives it has to check before (pre), during, or after (post) the execution of the elementary operation.

The second expansion phase is related to the diagnosis function. After detection of an exception, a diagnosis is performed to gain additional information on the nature of the exception and to update the internal model of the robot environment. For each of the elementary operations, a fault tree is modelled. Currently the fault tree models are static and do not depend on the structure of the robot program. The diagnosis activity results in the classification of the exception. In Table 25.2a the possible exceptions are listed. Each type of exception is further divided into several variants, depending on the results of the diagnosis.

COLLISION	- Collision with unknown object (C)
	- Collision with unknown object at position P (CP)
	- Colision with object O at position P (COP)
OBJECT ERROR	- Object lost at unknown position (O)
	- Object lost at postion P (OP)
NOT REACHED GOAL	- not reached planned goal, now at position P (CE)
OBSTRUCTION BY FORCE	
	- Obstruction with unknown object (F)
	- Obstruction with unknown object at position P (FP)
	- Obstruction with object O at position P (FOP)
GRIPPER ERROR	- Object moved in gripper (GE)

'PlanPickUpPart',
'PlanGetNewPart',
'PlanPathPassingO',
'PlanPathPassingP',
'PlanMoveOver',
'PlanNewGripPos',
'PlanNewUnGripPos',
'PlanObjectCorr',
'PlanCallibration',
'PlanRetry'

Tables 25.2-a, -b : Exceptions and recovery planning strategies.

Knowledge of the task structure of a robot program enables two approaches to exception handling: recovery planning and task rescheduling. In the first approach, an attempt is made to recover from the exception by replanning the robot actions. For each EO and the possible exceptions, a set of strategies is defined. Each strategy lists a number of motion or grasp planning activities which are executed one by one. In this way a detailed action plan is built which is then executed by the robot. In total 10 strategies were used to deal with the exceptions of Table 25.2-a.

If the exception was not identified, or if it was impossible to generate a recovery plan from the selected handling strategy, the robot cannot perform its production task successfully. The remaining elementary operations of the task are skipped and the exception handler now applies the scheduling scheme to find a subtask which is not affected by the exception. As described in section 17.3.2, the rescheduler uses a cost function to find a suitable sequence of subtasks. The cost function is also used to express the impact of the exception on the rescheduling activity. To do so, a so-called rescheduling strategy is used. For each subtask/exception pair a strategy is specified which indicates how the cost function has to be changed. The default strategy used in the kernel is to skip the current subtask and look for the first executable subtasks according to the order in which the tasks are represented in the robot program. In a realistic application environment the application specific conditions need to be taken into account as well. This activity is performed by the application programmer in the next programming phase.

II) The objective of this programming phase is to enable an application programmer to tailor the exception handling behaviour of the expanded robot program to the specific needs of the application. Knowing the robot which is selected, the configuration of the robot and peripheral equipment and process parameters, it is now possible to evaluate the functionality of the exception handling kernel and apply changes if needed.

The simulation system offers tools for testing and validating the exception handling capabilities of developed robot programs. In particular, the association between exceptions and applicable recovery or rescheduling strategies can be evaluated by the programmer. The application programmer can interactively tailor the available strategies to the needs of the application or add new strategies to the system. During the execution of the robot programs and the handling of exceptions, the application programmer requires information on the functioning of the HLI. The simulation module presents the status of the HLI and reports the results of the monitoring and diagnosis activities. Also a trace is presented of the reasoning process which is performed to handle exceptions. This trace provides information on the functionality of the exception handling strategies and thus enables the user to alter the strategies when needed.

III) The objective of this module is to offer tools for the generation of robot programs suitable for execution on existing robot control systems. For instance the generation of code for a Bosch controller is carried out. From the internal format of the elementary operations, the executable robot instructions are generated in BAPS (programming language of Bosch robot controllers). During execution of the expanded robot program, each elementary operation is translated to the corresponding BAPS expression which is then downloaded to the robot. The control of the program flow remains with the HLI. This way the exception handling mechanisms of the HLI can be fully exploited. A consequence of this approach is that the program generation module is needed both off-line for simulation of the robot and on-line for controlling the real robot.

25.2.4 Realization

For the realization of the HLI the programming language Prolog was used. The implementation of the modules is structured according to the functional decomposition of the HLI shown in Fig. 25.6. The modules described in 25.2.2 form the CECOSI system (CEll COntrol SImulator), the modules described in 25.2.3 form the programming system called REXSIM (Robot EXception SIMulator). CECOSI takes as input the assembly graph, the components functionality and capacity, the parts description and the user-generated part-specific schedule. Based on the internal representation of the cell in which the component states and transitions and the relations between the components were specified, the system offers a user-interactive process simulation in which the planned programs can be tested and verified. As output, a validated program, listing the elementary operations and the task information for each active component, was generated.

REXSIM reads in the program and expands the elementary operations with the exception handling kernel. In Fig. 25.11 a part of an interface file is displayed.

```
start_task(assemble_part,[sideplate]),
start_task(get_part,[sideplate]),
    transfer([],plat1pos3_appr,plat1pos1_appr,3,kamro_arm),
    approach([],plat1pos1_appr,plat1pos1,1,kamro_arm),
    grasp(sideplate,1,kamro_arm),
    depart(sideplate,plat1pos1,plat1pos1_appr,1,kamro_arm),
end_task(get_part,[sideplate]),
start_task(insert_part,[sideplate]),
    transfer(sideplate,plat1pos1_appr,pos01_appr,3,kamro_arm),
    insert(sideplate,pos01_appr,pos01,1,kamro_arm),
    detach(sideplate,1,kamro_arm),
    depart([],pos01,pos01_appr,1,kamro_arm),
end_task(insert_part,[sideplate]),
end_task(assemble_part,[sideplate]),
```

Fig. 25.11   Section of interface program in prolog notation containing task statements and elementary operations.

To simulate the execution of the elementary operations and to provide the application programmer with adequate information, REXSIM executes a logical simulation of each elementary operation. The reasoning process of the HLI is shown and interaction is enabled at each decision taken by the HLI. Any changes made by the programmer are automatically stored with the corresponding elementary operation.

```
  eo(3,eo_transfer,[[],plat1pos3_appr,plat1pos1_appr,3,kamro_arm],
[ [                           ('Object           lost          at          P',
              ['PlanPickUpPart',        'PlanGetNewPart']),
                              ('Object              lost',
    ['PlanGetNewPart']),
                    ('Collision          O       at          P',
    ['PlanPathPassingO','PlanPathPassingP','PlanMoveOver','PlanRetry']),
                          ('Collision            at          P',
          ['PlanPathPassingP',      'PlanMoveOver',      'PlanRetry']),
    ('Collision',
                        ['PlanMoveOver',       'PlanRetry']),
                          ('Gripper            Error',
    ['PlanObjectCorr', 'PlanGetNewPart'])]]
```

Fig. 25.12 Section of expanded output program for elementary operation 'transfer'.

In addition to the logical simulation, a geometrical simulation of the elementary operation is possible. The HLI can be connected with a graphical simulation system (ROSI) to display the robotized cell in full geometrical details.

The prototype was implemented on a SUN/Unix platform. The Prolog interpreter offered adequate language interfaces, as well as interfaces to the window oriented operating system. The graphical simulation system was a Unix implementation of ROSI (originally developed for VMS systems). The graphical interface made use of the PHIGS graphics standard.

25.3 Conclusions

The application modules of the prototype showed the feasibility of advanced concepts in different hierarchical levels of control. On the cell control level it was shown that, based on the specified functional cell model it is possible to plan and simulate the execution of different tasks. The synchronization of components and the time-parallel execution of operations were controlled by the HLI based on the functional cell model. On the robot control level the task plan was augmented with automatically generated monitoring and recovery procedures. During simulation the user could generate exceptions to test and verify the control procedures. Based on generically formulated recovery functions it was shown that task rescheduling mechanisms can be used to ensure continuation of an assembly process. Knowledge of the application and system configuration can be used to improve the efficiency of the exception handling capabilities. Both for rescheduling and recovery planning, the system requires a description of the task structure of the robot program. Future work will concentrate on the formal representation of this task structure and on the definition of task primitives to extend the use of rescheduling strategies.

The two-level HLI implementation of CECOSI and REXSIM was connected through the information system, which enabled the use of a uniform and powerful representation scheme.

Another result was the design of common data base structures in order to exchange information between this system and other systems. Cell structure and layout, task description and product parameters are exchangeable via Rdb structures.

Chapter 26

Industrial applications of the planning and interactive programming system

B. Duffau
Renault Automation

26.1 Overview

26.1.1 General presentation of the demonstrator

The Planning and Interactive Programming demonstrator is oriented to industrial job type actions, such as welding, glueing, laser cutting, assembly, or pick and place. It is mainly composed of two integrated subsystems, the planning subsystem and the explicit programming subsystem.

The purpose of the planning subsystem is to help an industrial designer who must perform a robot application to build the tool trajectory and to describe the robot cell layout. To be as automated and fast as possible, the planning work is aided by expert systems, corresponding to the target application job type.

The purpose of the programming subsystem is to adapt the tool trajectory to the selected robot, to simulate and improve the trajectory, and translate it into the robot controller language.

Two interconnected programming subsystems have been developed:

- R.O.S.I. (RObot SImulator) which has been built by the University of Karlsruhe, is supported by an explicit programming language.
- R.I.E.P. (Robot Interactive Explicit Programming) which has been built by Renault Automation, is 'interactive', i.e. the rough trajectory coming from the

planning subsystem can be modified interactively by the designer with a mouse or directly improved by algorithms.

One of the major goals of this demonstrator was to be independent from the host computer database structure as far as possible. To reach this goal a concept of semantic robot cell structure and a neutral database interface have been developed and tested; the ROSI subsystem has been implemented on the DEC-RdB database system, and the RIEP subsystem has been implemented on ORACLE database. Through this neutral interface, the two systems exchange robot cell information and robot trajectory in a semantic way (i.e. according to the meaning of the information, and independent of the syntactic database implementation of the information.)

26.1.2   The planning subsystem

The purpose of the planning subsystem is to help and lead the designer from the application description to obtaining a trajectory in terms of tool movements.

Fig.26.1 shows globally the planning steps: according to the manufacturing problem definition and an elementary cell description, two expert systems will help the designer to obtain a detailed cell layout (including geometric frames) and a semantic description of the tool actions.

These tool actions will be automatically translated into the actual tool trajectory and into an application folder which will keep track of the application planning approach, for the purpose of maintenance.

The two expert systems are:

(i)   A tool actions expert system:this helps the designer to split the global task which must be performed into various levels of detailed actions up to elementary actions which correspond to trajectory elements. The knowledge which is contained in action trees, and which is kept in the application folder, can be used to extend this action to variant trajectories or to generate a set of various trajectories.

During the splitting of the actions, the expert system analyses the cell elements for which it needs more information (such as catching frames of parts) and will use rules to determine them, or ask the user to give them, through the CAD display.

(ii)   A methodological actions expert system: this leads the designer to the actions which must to be performed to satisfy the applications requirements. Methodological actions and requirements are interconnected in a network. One of

285

the main benefits of this approach is that the network may be built by the expert, and that this network is completely transparent for the designer, i.e. as the planning system has been fully integrated into a CAD (Euclid) system, the designer only sees a menu. The variations in the 'intelligent' menu items are automatically performed by the expert system.

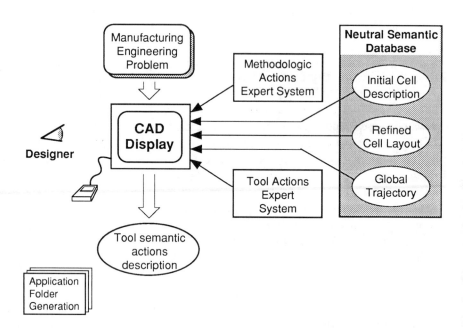

Fig. 26.1   Planning subsystem.

26.1.3   Programming

After the planning steps, the system knows the trajectory in terms of tool actions. It is now possible to choose the robot (if it has not already been done). Algorithms can define the robot position in the cell. Generally the target position is in a horizontal 2-D area, but in some special cases it is necessary to install the robot hanging from the ceiling or in another complex position.

Automatically, the tool trajectory is transformed into a robot trajectory, if every geometric frame is reachable.

During the programming steps, graphic simulation and interactive facilities allow the designer to improve the robot trajectory. This improvement can also be done through optimization algorithms.

When the trajectory is finalized, a postprocessing of the module translates the trajectory into the language of the robot controller, and after calibration the robot is ready to operate.

26.2 Industrial application: Sheet metal welding

Previously, spot welding points were determined on a part plan, then trajectories were obtained by piloting the robot by hand with the buttons board of the controller.

Two main difficulties with this approach were the impossibility of trajectory improvement and the risk of collision which obliged the approach to the part to be made in very slow motion.

The trajectory learning was therefore a time-consuming operation.

The new approach consists in the following methodology (planning, then interactive programming):

- the part is extracted from the CAD database;
- the welding points are determined in a 2 D view;
- the welding spots are automatically projected onto the CAD part and the normals are evaluated;
- the computer proposes references (geometric frames) for the grip approach;
- grips are proposed and the interferences between grip and parts are evaluated,
- the robot is selected and automatically positioned so that all the points may be reached;
- a first trajectory is implicitly computed and simulated;
- the trajectory is translated into the robot language;
- the actual cell is calibrated, and the robot operates.

These actions have been performed on different benchmarks, and paticularly for the assembly of elements on car bodies. Spot welding benchmarks have been performed for automotive industry subcontractors.

The reduction in time from the previous method to the new method varies from one bench to the other. In certain examples, work which needed two days previously is now performed in two hours.

287

26.3   Industrial application: Laser cutting

26.3.1   Laser cutting

A laser cutting methodology has also been developed. It is composed of:

- a curve definition on the part,
- a curve discretization, according to different criteria,
- the laser trajectory definition, including the laser torch adjustment and speed control, particularly in the windings of the laser path,
- the choice and positioning of the robot,
- the simulation and trajectory improvement.

This method is in validation, particularly on sheet metal laser cutting applications for the cutting of holes in pre-production cars, where it is difficult to build specific dies.

26.3.2   A new technology: 3D laser cutting on short batches of complex shapes.

The laser cutting benchmark has an influence outside the automotive industry. In short batch production there are a lot of products which need various holes, different from one variant of a part to another. For this reason, such holes must be done at the final step of the manufacturing process, and are generally processed by hand. A laser is sometimes used, but it needs the part to be flat (2 D curve), and the shape of the hole to be simple.

Now it will be possible to use this technology for complex parts, and for complex shapes of hole, when the part has been represented in the CAD product design system. The research is now progressing in two ways:

- the first consists in determining what kind of product may use that technology (for example: bath tubs, yachts hull...)
- the second consists in extending this approach to another cutting system: the high pressure water jet cutting.

288

Fig. 26.2   Laser cutting.

26.4   Industrial application: Assembly of bulky parts

This application has been studied for the automated assembly of seats into cars and is in study for the assembly of dashboards.

The methodology helps to solve two categories of problems in such assemblies:

- first the automated introduction of a bulky part, such as a seat or a dashboard, into the body of the car, without collision;
- then the difficult positioning and tilting movements.

Fig. 26.3   Assembly of seats.

26.5   Application in the nuclear industry (hardfacing)

In the nuclear industry, some nuclear tanks are built in steel and covered with stainless steel by hardfacing.

Such complex tanks are composed of pipes, cylindrical shells, semi-spheric shells, which may be hardfaced with machines.

The problem is at the intersection of shells and pipes where there are various radii of curvature in connection. No machine can perform the operation.

Previously, the operation was done by hand, at a temperature of 200° C, and with a high quality requirement.

The bench has been done with a robot ACMA Y28, hanging from the ceiling of the cell.

Another requirement for this problem was as follows: at each pass, the width of the stainless stripe is constant, and the trajectories must be at the same curved distance from one pass to the next to avoid overlap of specific 'parallel' trajectories on complex surfaces.

In this example, 6000 hours of painful work have been replaced by 2500 hours of robot work, with consistent quality.

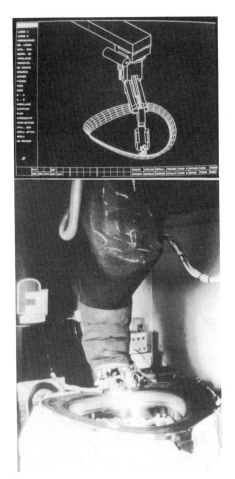

Fig.26.4   Hardfacing.

26.6   Application in the aerospace industry

In the aerospace industry, tests are in preparation for the trajectory design of a manipulator's arms with an important number of degrees of freedom. This manipulator was designed to the inside an aerospace shuttle. When the upper door opens, the manipulator expands its arms out of the shuttle and performs maintenance operations.

26.7 Conclusion

The first prototypes were developed in the automotive industry, but applications have appeared in other industrial domains such as the nuclear industry and the aerospace industry, and it seems that extensions may be done in a lot of traditional short batch production industries, where the complex shape of parts previously prevented the use of laser cutting or high-pressure water cutting. With the interaction of the planning expert system and the interactive robot programming, the design system may be accessed in terms of designer's needs, corresponding to various industrial job type methodologies.

Chapter 27

Applications at FIAR

A. Pezzinga
FIAR Spa., Italy

The results and benefits of the ESPRIT 623 project, for FIAR, apart from the acquisition of know-how, of personnel and of hardware and software purchased during the project, consist mainly in the realization of a prototype expert system called IMPRES (IMplicit PROgramming Expert System). The goal of this system is to speed up the programming activity of a robotic workcell for assembly tasks, allowing to execute it off-line, by means of advanced planning and scheduling capabilities.

27.1   Prototype description

27.1.1   Inputs

The system receives as input the 'task description', and the 'contact definition'.

The 'task description' gives the initial (before the assembly) and final (after assembly) positions of all the workpieces. This information completely describes the assembly task. This task description can be quite easily obtained using an interactive CAD-like system, able to define relative positions of objects.

The 'contact definition' specifies the type of the contacts between objects in the assembled configuration (e.g. screwing, insertion,...).

The system also makes use of a World Model built from the information given by the user, which contains:

- The geometrical description of all the elements involved (workpieces, fixtures, robot links,...), obtained through a standard interface (FEMGEN) provided by many CAD systems.

- The model of the robots (kinematic parameters, definition of the robot language).

- The cell layout (robots and fixtures positions).

27.1.2 Function

From the input information the system is able to find the executable program for the robots of the cell.

The programming activity is divided into three sequential phases:

1. ACTION PLANNING: in this phase (independent from the robots) the assembly constraints are exploited in order to obtain a partial ordering of the 'actions' (an action is the operation required to assemble a single workpiece). Please note the fact that the user is not required to define the order in which the various parts have to be assembled.

   The constraints are generated by geometrical considerations (e.g., the assembly of a workpiece must not obstruct the subsequent assemblies) and by stability considerations (a workpiece must be stable in the final position).

   Moreover, the parameters of each assembly action (assembly, direction, departure,...) are computed.

   The action planner generates as output an assembly graph, in which the partial ordering of the assembly operation is made explicit.

2. SCHEDULING AND MOTION PLANNING: in this phase (using the description of the robot geometry and kinematics):

   - each action is assigned to a robot;
   - the actions are completely ordered;
   - the collision-free trajectories of the robots are computed;
   - the synchronizations between the robots are defined.

   These activities take into account further constraints (working area of the robots, collisions between robots) and try to optimize the overall assembly time.

   As output of this phase, a completely defined program for the cell is obtained, in an internal high-level language.

3. PROGRAM GENERATION: in this phase the executable program in the syntax of the target robot language is generated.

27.1.3 Output

The output of the system is the executable robot program.

27.2 Action planner

The Action Planner is a two-level system which performs a simulation of the assembly to determine the ordering of the operations.

The lower level manages the geometrical information, detecting collisions between objects to find allowable assembly directions and to determine if an object is constrained in such a way as to be stable given the other previously assembled objects.

The high level is a rule-based system that performs the search of the assembly order exploiting the knowledge about the domain and the specific problem.

The system finds the assembly graph, that is the partial ordering of the assembly operations. The assembly graph is a sequence of steps, each being a list of names of parts that can in principle be assembled in parallel (whether this will be actually possible or not depends of course also on the robots' positions).

Having completed all the assemblies of one step, the assemblies of the following step can be executed without conflicts, that is, without instability of parts or collisions between them.

The Action Planner tries to maximize the parallelism of the assembly operations, that is, the number of assemblies belonging to the same step. A simulation of the assembly is performed according to the following control cycle:

1. The parts to be assembled next are chosen among all the workpieces that are in contact with at least one previously assembled workpiece. At the start, all the parts are not assembled, and the ones in contact with the assembly fixture are chosen.
   The system recognizes that a workpiece is in contact with another if there is an overlap between the surface delimiting them.

2. For each workpiece in the candidate set, the system checks if the final position is stable. This is done by checking if the contact surfaces constrain the workpiece (which is candidate for assembly) in the final position in such a way as to enable only movements with an upward component.
   If a workpiece is not stable in the final position, it is eliminated from the candidate set: for assembly it is likely that this workpiece will be made stable by some of the workpieces not yet assembled; therefore, its assembly is postponed.

3. For each workpiece in the candidate set, the system finds the allowable assembly directions. Because all the fine motions for mating workpieces are assumed to be

linear, these directions can be obtained examining the orientation of the contact surfaces.

These directions are then checked for collision: if during the approach to the assembled position a collision occurs with a workpiece already assembled, that direction for that assembly is eliminated.

Also the collisions due to the gripper are taken into account.

If an assembly is not possible in any direction, it means that one or more already assembled objects block that assembly. At this point the system tries to solve the problem using a set of strategies to reorder the plan.

Strategies try, for instance, to disassemble objects (remember that since the assembly is only simulated, this corresponds to altering the order in which the assemblies are performed) or to build separate subassemblies in different areas of the workplane.

The Action Planner applies the 'Goal Regression' planning techniques to reorder the assembly plan. If this technique fails, it applies another method, which we have called 'Goal Clustering', which results in the creation of separate subassemblies.

It is important to observe that the number of checks is not as large as it may appear: the number of allowable assembly directions is, in general, very small: in almost all the situations there is only one allowable direction.

4. All the remaining objects in the set are collected into a step of the assembly graph, as they can be assembled in parallel - without taking into account additional constraints due to the robots. The assembly operations are performed and the cycle is repeated until all the workpieces are assembled.

It is important to observe that, within the previous scheme, some very common assembly operations are not possible: for instance, the complete insertion of a pin into a hole in an object which always causes a collision between the fingers of the gripper and the object.

The definition of the type of each contact has been introduced precisely to take into account these situations: in this example, the contact between the pin and the object will be defined as 'complete insertion'.

The system has a set of rules in a knowledge base which define a strategy for solving the 'complete insertion problem': in this situation the assembly will be done in two steps: in the first one, the pin is inserted only partially into the hole, until the fingers are in contact with the object; in the second step, the fingers release the pin and push it completely into the hole.

This method gives knowledge to the system about a set of common techniques used in assembly.

Of course this set cannot be complete: but the system allows the user to modify its behaviour; the user can discharge operations which the system believes are possible; conversely, the user can resume operations unnecessarily eliminated by the system.

27.3   Scheduler

This module is also a two level system: the high level has the task of scheduling the operations to the robots of the cell, and the low level has the task of finding collision-free trajectories between the grasping positions and the releasing positions.

The low level (motion planner) is described in the next section. The purpose of the scheduler is to sequence the operations of the same step of the assembly graph and to assign them to the robots.

As has been described in Chapter 16 (Task Level Programming), usually the number A of operations in the same step is quite small (5-6 as a maximum) and the number R of robots is also small (2-3). Because each operation of a step must be performed before each operation of the following step, each step can be considered independently from the others. This is actually a simplification which reduces the complexity of the problem, reducing also the optimality of the solutions found.

A tree is built generating the possible assignments of operations to the robots; if a robot cannot perform an operation, because either the initial position or the final position of the workpiece is out of its working area, the assignment is discarded. A search for a pseudo-optimum path is performed on this tree, using the A* algorithm, estimating the time required for the transfer movements assuming linear trajectories (in joint-space) at maximum speed.

This is the function g of the algorithm (remember that the function which guides the expansion of the states is f = h + g), while h is the time to actually execute the assembly operations. This is obtained by activating the Motion Planner to compute the collision free trajectories for the robots. The function g is admissible, that is, it gives a value always less than or equal to the real one, because in general the trajectories are not linear, and not executable at maximum speed.

It can happen that the Motion Planner is not able to find a trajectory of this sequence, either because this trajectory does not exist or because it is too complex and a computational time limit is reached. In this case the Scheduler backtracks, choosing a different path on the tree. This choice is done trying to save as many trajectories as possible, because the motion planning is a quite time-consuming job.

Of course, the method presented works only when the number of operations that can be performed in parallel is small: if this is not true, the system asks the operator to divide the operations into subsets in order to keep the size of the search space small.

As is clear, there are a number of approximations of the scheduling solution found with respect to the optimal one. This happens because finding optimal scheduling solutions requires a complex interaction between scheduler and motion planner, with many backtrackings from the scheduling level and the motion planner level. Because both motion planning and optimal scheduling are time-consuming activities, an exact approach is not applicable, or a too high computational time will be obtained.

27.4   Motion planner

Many approaches have been studied in motion planning: the one more commonly used is the 'configuration space' approach. This approach is based on a precomputation of a suitable description of the free regions in the manipulator joint space; then a search in the connectivity graph of these regions is performed.

The main advantage of this method is that the search of the optimum path is fast, even for very cluttered environments, after the precomputation of the free regions. Unfortunately this precomputation is very complex.

If many trajectories have to be computed in the same environment, this precomputation can be done only once.

Unfortunately in an assembly task each time an object is assembled the situation changes, and the free space has to be recomputed.

FIAR motion planner adopts a different approach. It performs a weighted A* search in the configuration space, from the start to the final position (in joint coordinates).

The space is discretized into n-dimensional cells (for a robot with n degrees of freedom) with a given discretization factor. This space is not explicitly represented, but one can think of it as an n-dimensional grid, in which each cell represents a robot configuration in joint coordinates; each cell has a side length depending on the discretization factor.

The cell corresponds to a node of a graph in which the arcs represent adjacency relations.

The starting position and the final position are two attributes of the nodes: the system generates a search tree in graph, with the A* algorithm, using as cost function the distance in joint space. With this approach, the system finds the minimum length path in joint space.

The complexity of the search depends, of course, on the number of cells: this number depends on the number of degrees of freedom of the robot and on the selected discretization.

It can be noted that this method is similar to the configuration space one, with two major differences:

- the whole configuration space is not generated, but only the part required in the search;

- the forbidden space representation is simply the list of the cells in which a collision occurs.

The system works at two levels: the first level finds a preliminary path with a rough discretization; the second one refines this path, finding – with the same method - the subpaths between the intermediate positions of the first level path. At both levels the search of the path can fail, either because a collision-free path does not exist or it is too complex to find. A failure at the second level is reported to the first one, allowing backtracking between the levels. A failure at the first level is reported to the scheduler and a different scheduling strategy must be attempted.

This approach works when the cell contains only one robot. In a cell with more than one robot some extensions are required. Our solution is based on a two-level collsion avoidance between robots:

- The high level simply keeps each robot away from others, with synchronization techniques: in a cell with more than one robot, there are regions reserved to one robot (for instance, each robot has its own pallet from which to grasp the workpieces) and shared regions (for instance, the assembly area). The high level assures that only one robot at a time is within a shared region, by means of explicit synchronization. This high level is implemented in the scheduler; it is presented in this section only for clarity reasons.

- The low level plans the trajectory for one robot at a time, taking into account the current position of the other robots.

27.5   Program generator

After the scheduling and motion planning activity the system has produced a high level program for the cell in which all the execution details are defined.

The Program Generator, using a knowledge base describing the target robot language from both a syntactical and a functional point of view, translates this high-level program into the executable robot program.

The knowledge base representing the robot language can be viewed as a simple kind of semantic network: a node represents a type of operation (e.g. 'grasp', 'assemble', etc., at a high semantic level; 'open_gripper', 'depart', 'move_linear' at target language level) with slots defining the parameters of the operation (e.g., a trajectory,. a position, ...).

The nodes connected with the one-to-one relation 'is_a' (e.g. 'screwing' is a kind of insertion) and the one-to-many ordered relation 'is_decomposed_into' (e.g. 'grasp' is decomposed into 'open_gripper' 'approach', 'close_gripper', 'verify', 'depart').

In this way the high-level operations are defined in terms of more simple operations; this decomposition process continues until a definition in terms of robot level primitive operations is reached.

Moreover, the is_a relation allows the high level operations to be specialized into subtypes in order to manage the special type of contacts used in the Action Planner, at this level also.

Inheritance mechanisms and procedural slots allows the parameters of a low-level operation to be differentiated from the parameters of a high-level one.

The program generator is written in a rule-based language (OPS83); in order to have the described behaviour, we have implemented a knowledge compiler that reads the description of the target language and translates it into a set of OPS83 rules; these rules are then compiled and linked with the program generator.

27.6   System validation

The system has been tested on a set of examples. Among them, one is based on the 'Cranfield Benchmark' assembly. This benchmark has been defined mainly as a test for robot accuracy; it has been chosen because it is representative of a quite large class of assembly tasks.

Our example task requires 14 workpieces to be assembled, using two 6 degrees of freedom PUMA260 robots, a fixture for the assembly and two pallets for the workpieces. Within this example, the geometry is quite complex (about 2000 faces for describing workpieces and fixtures and 1000 faces for describing each robot). Conversely, the logical structure of the assembly is quite simple – in the Action Planner no backtracking is required.

The CPU time required to obtain the program – on a DEC VAX Station – is less than 17 minutes, excluding time for graphic presentation and files reading and writing. The complete execution, including graphical simulation of the various planning phases, requires about 75 minutes.

This is clearly quite a good result; similar performances can be expected from many assembly tasks.

More 'difficult' assemblies might require human intervention or a greater computing time; the solution of this problem, based on the experience we have made, relies in our opionion on a high degree of interactivity between the system and the user.

27.7   Code dimension

The size of the described system is 33,300 lines of source code; 15,500 lines are Pascal code (for geometric computation, CAD interface and motion planning) and 17,800 lines are OPS83 code (for the inferential parts).

A large part of the Pascal code – collision detection, geometric model management – is used by both the Action Planner and the Motion Planner.

The present version of the system is only a prototype; some increase in the system size is foreseeable – mainly in order to have a better user interface.

300

In any case, the above values allow the estimation of the typical dimension of the kernel of a helpful task-level robot programming tool.

27.8   Future work

Although IMPRES has been developed for the industrial field, its main applications have been so far in Space Robotics. FIAR has been active in space since the early sixties, mainly with contracts from the European Space Agency (ESA) and the Italian Space Agency (ASI).

By means of a cooperation with the Italian company Tecnospazio, the adaption and delivery to ESA of a number of products and software modules which have been born as a follow on of IMPRES has been agreed.

The application of software modules developed for industrial applications (or maybe one should say 'for earth applications') requires, of course, various adaptations. An example for all are the different strategies and techniques which have to be followed in space for robotic assembly, in the absence of gravity, compared to similar on-ground applications, where gravity can be exploited for instance for 'stability of assemblies' considerations.

The kernel of the system developed in ESPRIT 623, however, remains valid, reducing the development effort by a large degree.

The main applications of IMPRES, or modules of it, to the space robotics field are the following:

- ARCS (Autonomous Robot Control and Simulation): for the study of control architectures for autonomous or semi-autonomous robots in space, including planning, sensory data fusion, replanning and simulation. This project, started by ESTEC (European Space Technology Center), one of the ESA centers in Europe, is a direct spin-off of ESPRIT 623.

- STEREOVISION: started by ESTEC and ASI to study the possibilities of using stereovision information, obtained through a pair of cameras, to guide a robot for internal space applications. In this project the high-level programming interface of the system, which uses the interpretation of the sensory information determined by another expert system to determine the program to, for instance, grasp an object, is an adaption of the modules Action Planner and Program Generator of IMPRES.

Chapter 28

Applications at KUKA and benefits

P. Weigele
KUKA Schweissanlagen & Roboter GmbH

KUKA Schweissanlagen & Roboter GmbH is a manufacturer of industrial robots and complete user solutions. The optimum integration of industrial robots into complex manufacturing systems which calls for highly qualified engineers is a decisive factor in the efficient operation of a system.

This engineering work can be supported by computer-aided tools, especially robot simulation. The software package developed by KUKA is an integrated system for planning, design and programming of robot cells. In the following two sections typical applications managed with the system are described. In the final section the benefits of the system are summarized.

28.1 Application: Palletizing of clutch cases

The clutch cases arrive at a conveyor. The robot has to pick the cases with a special two-arm gripper from the conveyor and place them successively in a rack (see Fig. 28.1).

By means of the planning system the following problems are to be solved:

- What is the optimal component selection?
- What is the optimal arrangement of the components within the cell?
- Can all target positions on the pallet be reached by the robot?
- Does the robot handling make the work piece collide with other cell components?

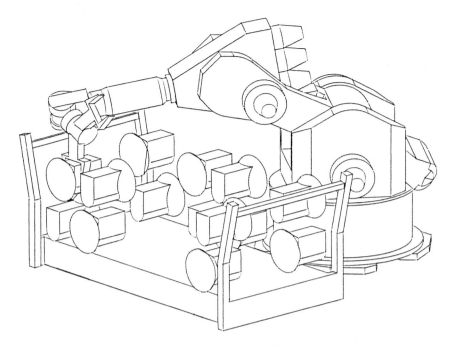

Fig. 28.1   Task execution in simulation.

Reachability of all positions in the rack and collision avoidance are the main problems in this application because of the restricted spatial conditions. The above-mentioned questions cannot be solved using conventional methods. Analysis of the problem at the drawing board is not possible because of the 3-dimensional nature of the problem. Workshop tests cannot be performed because the iterative solution of the problem requires frequent change of the arrangement of the components.

The robot cell is designed, programmed and analysed with the system developed by KUKA. The following steps are performed:

In a first step, the required cell components are selected. Standard components are maintained by a relational database management system.

A KUKA IR161/60 robot with 400mm arm extension is loaded from the robot library into the cell. The robot tool (the two-arm gripper), the workpiece (clutch cases) and the peripherals (rack and conveyor) are modelled with the solid CAD modeller integrated in the system.

The preliminary layout of the cell is defined. Restrictions made by the customer have to be taken into account.

The target positions of the clutch cases within the rack are set by means of the CAD modeller. As they are defined relative to the rack position, subsequent movements of the rack affect these positions in the same way. In the next step the action sequence is defined. The main problem is now to find a robot location where all target positions in the rack can be reached by the robot without collision.

An optimization utility is invoked to provide the engineer with suggestions on how to position the robot. To reduce computation time and response time for the user of the system, the optimization procedure can be influenced in several ways. The variant parameters can be selected from a parameter set which depends on the chosen optimization procedure. The parameter range can be limited. Continuous parameter spaces can be approximated by equidistant discrete values.

The system successively provides robot positions where the robot program can be executed without regard to collisions. They must be detected by the user of the system.

In this case none of the suggestions made leads to a satisfactory robot program. This is not due to insufficiency of the optimization procedure but results from the complex nature of the problem. But during the interactive optimization process, the engineer gets hints from the system where promising improvements can be made. Collisions caused by the geometry of the gripper turn out to be the main problem. Therefore the gripper must be redesigned. With the modified gripper, a robot position can be found where all positions in the rack can be accessed without collision.

With the results achieved during simulation, the real robot cell can be designed and delivered to the customer (see Fig. 28.2).

Fig. 28.2 Task execution at the real system.

28.2 Application: Seam welding of motor carrier

A motor carrier is clamped and positioned on a turntable with a fixture. Two robots have to carry out stitch welding and arc welding of 112 linear and circular weld seams (see Fig. 28.3).
By means of the planning system the following problems are to be solved:

- What is the optimal component selection?
- What is the optimal arrangement of the components within the cell?
- What is the best partition of the welds? Which robot welds which seams?
- Can all welds be reached by the robots?
- Are there collisions between the robots and other cell components?
- Does the estimated cycle time meet the demands of the customer?

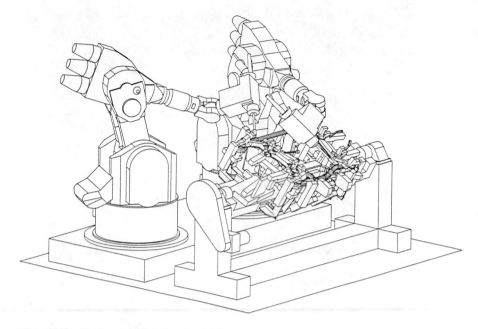

Fig. 28.3   Task execution in simulation.

Additionally, motion programs for the RCM control in SRCL language (Structured Robot Control Language) are to be generated.

The main problems in this application are accessibility of all welds, collision avoidance of the robots and the control of the technological welding process. The robot cell is designed, programmed and analysed with the KUKA system. The following steps are performed:

In the first step the required cell components are selected: Two KUKA IR161/15 robots with 200mm arm extension and three different welding torches which seem suitable for the application are loaded from libraries into the cell. The motor carrier is provided by the customer and loaded into the cell using the neutral VDA-FS interface. The fixture and the turntable are modelled with the solid CAD modeller.

The preliminary layout of the cell is defined. Restrictions made by the customer have to be taken into consideration. An initial position of the turntable which remains fixed during the welding operation is defined.

The geometry of the welds is defined relative to the workpiece. The start and end of the seams are marked by target position set by means of the CAD modeller. The shape of the seams is composed of linear and circular curve segments.

In the next step, the welds are optimised towards requirements of the technological welding process. The system offers features to optimise the orientation of the workpiece in space as well as the orientation of the welding torch relative to the seams.

For the definition of the program, robot-specific aspects must be considered. In the case of two robots welding the same workpiece, the welds are partitioned in such a way that collisions between the robots will probably be avoided. In the last step the sequence of actions is planned for each robot.

The result of the planning procedure is a first version of the robot program. It can be executed within the test environment. The robot movements can be visualised on the graphic screen in real-time.

Usually the first version of the program does not meet the demands. The experienced user has to use his expert knowledge and perform the preceding steps iteratively to elaborate the robot programs. He is supported by analysis, e.g. cycle time, or peak velocity of an axis provided by the system.

In this case, the following modifications have to be made to grant accessibility and collision-free paths:

- change of welding torch;
- repositioning of components within the cell; and
- change of partition of welds between the robot.

It turns out that, even with optimal component selection and layout some of the required welds cannot be accessed by the robots. After consulting with the customer this can be accepted.

A postprocessor is invoked to compile the programs developed with the planning tools to the native language of the robot control unit.

The real system is built according to the optimised layout. Deviations between the workpiece position in the CAD model and in reality are automatically compensated for by modifying the compiled programs systematically. This is done by a PC-based system that also transfers the programs to the robot control. Finally, the program is tested in the real system (see Fig. 28.4).

Fig. 28.4   Task execution in the real system.

28.3   Benefits

Since 1986 the system has been established step by step in the areas of planning, mechanical design and programming. Five workstations are currently installed at the KUKA site. In the pie chart shown in Fig. 28.5 two different application areas within the company are selected as an example to show the increasing importance of the system.

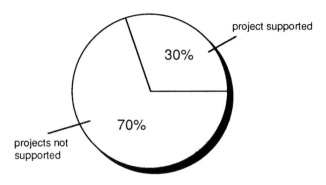

transfer lines / assembly lines
project supported by the system

project supported

30%

70%

projects not
supported

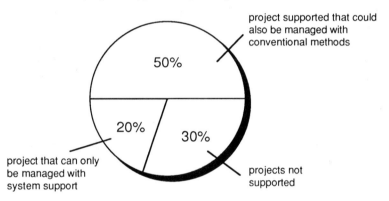

flexible manufacturing cells
project supported by the system

project supported that could
also be managed with
conventional methods

50%

20%   30%

project that can only
be managed with
system support

projects not
supported

Fig. 28.5   Projects supported by the system.

In the area 'transfer lines/assembly lines', one third of the stations are planned and designed using robot simulation. In the area 'flexible manufacturing cells' even 70% of the projects are supported by the advanced planning tools provided by the system. Due to the complexity of the applications, only 20% of these projects can be managed using robot simulation.

The advantages resulting from the application of the system can be summarized as follows:

1. Cost reduction

Personnel costs for design, optimization and programming are saved and costs for workshop tests and subsequent alterations to the system are significantly reduced.

2. Faster results

At a very early stage in a project, highly accurate information is available, e.g. regarding the configuration of the layout.

3. Time saving

User solutions are developed faster, also within the tender preparation phase.

4. Optimum engineering product

Several alternative solutions can be analysed and compared. This results in a decisive improvement in the quality of a solution.

The most important advantages are time saving and faster results. In the block diagram shown in Fig. 28.6, the average time spent in managing different application types with the planning system is listed in comparison to conventional methods.

For spot welding applications with five-axis robots, the time spent is reduced by 90%. Reachability of the target positions can be checked and the optimum robot location can easily be found by applying the optimization algorithms embedded in the system.

time expense for applications

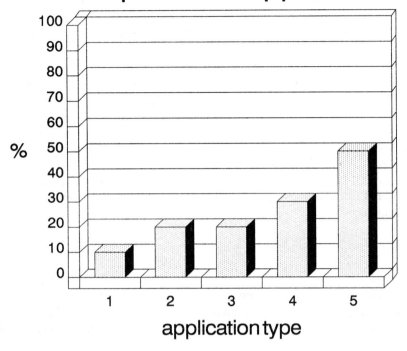

1   spot welding - 5 axis robot
2   handling applications
3   laser applications
4   spot welding - 6 axis robot
5   arc welding

Fig. 28.6   Time spent for applications.

Ambitious laser applications can often only be managed by using simulation. In those cases which can also be managed with conventional methods the time needed is reduced considerably. In operations like laser cutting, frequently complex kinematic structures such as flexible beam guidance sytems with a stationary laser beam are used. The demand for reachability of all target positions by both kinematic systems and for collision-free paths requires frequent modificaion of the arrangement of the cell components. This can easily be done using the optimization algorithms implemented in the system, whereas the realisation with workshop tests is often impossible.

For handling applications like machine tending and palletizing, often stringent time requirements have to be met. By means of the integrated cycle time estimator the time needed for the whole robot program and single motions can be calculated. Time-consuming critical motions can be optimised. Time saving is approximately 80%.

The results achieved with the system when managing spot welding with six-axis robots can be compared with the above-mentioned spot welding with five-axis robots. Additionally one degree of freedom, that is the rotation about the main axis of the tool, can be optimised.

When applying the system to arc welding applications, up to half of the time needed can be saved. Nevertheless, the projects suitable for simulation must be carefully selected. That is because the optimization of arc welding programs in many cases does not require mere motion planning but also the consideration of the technological welding process. This process can often only be controlled in workshop tests.

28.4   Conclusion

The above statements prove that the system developed by KUKA now influences all product divisions of the company. Time and cost saving, faster results and better product quality are the main advantages of the system. Due to the increasing complexity of the products, many projects can only be managed by support of the planning tool.

Therefore the course that is being pursued will be continued in future with the aim of further improving the functional scope of planning instruments and their integration into CIM structures.

Chapter 29

Applications at the IPK Berlin

G. Schreck, R. Bernhardt
IPK Berlin

29.1 Introduction

In the frame of ESPRIT 623, two main departments of the Fraunhofer Institute for Production Systems and Design Technology (IPK), Berlin, have been involved. These are the departments of robot system technology and planning technology. The following description of applications is mainly focussed on the activities of the first one in the field of explicit off-line programming.

During the ESPRIT 623 project, reference models for the structure of an off-line programming system were worked out by all partners. This work has led not only to common models, but also to partner-specific prototypes or testbed realizations. In the following, the IPK-specific approach of an off-line programming system is briefly explained. Furthermore, applications within the project and also adaptions to the requirements of other projects in the field of underwater and space are outlined.

29.2 Realized off-line programming system

The off-line programming system developed at the IPK Berlin, which is primarily a simulation system, is a design tool allowing the creation of application programs for existing or planned manufacturing systems. By transferring the application programming task to the planning area which precedes the actual manufacturing, the utilization of all data systems within a company (e.g. CAD systems, technology data banks etc.) is effected and thereby a continuous data flow from design to manufacturing is rendered possible.

The basic structure of the realized off-line programming system is shown in Fig. 29.1. It consists of the main components
- programming;
- simulation; and
- code generation and transfer.

The off-line programming system is based on the approach of a step-by-step programming sequence involving different planning levels (programming phases), whereby a continuous information flow is guaranteed (Fig. 29.2).

These system components are accessed via a dialogue-oriented user interface which also offers graphical I/O functions.

The system is characterized by a task-oriented execution of the programming task made possible by a user interface adapted to manufacturing tasks for which the necessary system functions are provided. In a dialogue with the system the user defines all geometrical and technological data describing the execution of the manufacturing task. Geometrical data are derived from the shape models of work pieces, tools etc. by CAD-functions. Technological data are supplied by the user with the help of pre-defined screen forms.

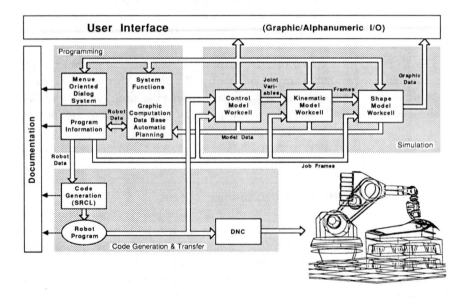

Fig. 29.1   Structure of the realized off-line programming system.

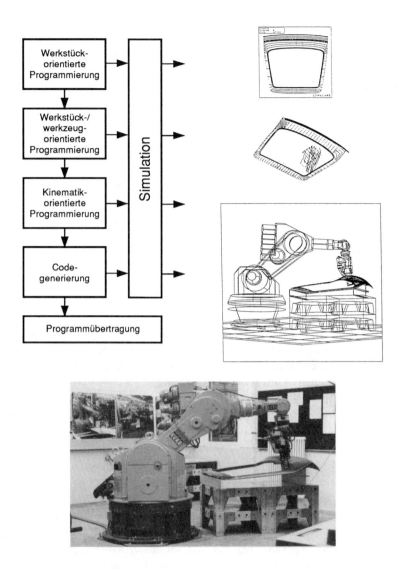

Fig. 29.2   Programming operation.

All information accumulated up to this point is integrated within the data structure 'program information' which is still independent from the robot controller to be used. In the subsequent code generation step, a robot program in

315

the language of the robot controller or an ESR program can be generated. The robot programs can be transferred to the actual controller via a DNC interface.

For programming as well as for testing the robot programs, simulation functions are available. They allow on the one hand a direct check of the user's inputs; on the other hand, high quality of the robot programs concerning their executability by the real manufacturing system is attained. For the simulation system computer internal models are necessary which reproduce the components involved in the manufacturing task with regard to their motion behaviour (control model), kinematic (kinematic model) and their shape (shape model).

As shown in Fig. 29.2, the programming process is to be executed in three parts. With workpiece-oriented programming, the task geometry is defined on the basis of shape models provided by a CAD system using graphical interactive functions or calculation functions. In the case of spot welding, this corresponds to the definition of welding spots on the used sheet metal parts.

In the case of workpiece/tool-oriented programming, the tool configuration necessary for correct task execution is done in relation to the workpiece. In the spot welding process this corresponds to the configuration of the welding gun at each welding spot. Furthermore, all technological data essential for the welding process have to be put in at this stage via screen forms.

In the case of kinematically oriented programming, the kinematics of the robot itself as well as its kinematic linkage with the work cell is taken into account. Due to kinematic frame conditions, or in order to avoid collisions, additional auxiliary motions can be defined in this programming phase.

Data input during the first three programming phases is stored by the system in so-called 'job frames'. In a dialogue this information is extended by the execution parameters (motion parameters, peripheral instructions and sequence) and finally integrated into the data structure 'program information'. In the code generation step the program information is converted into a robot program in the language of the actual controller or into an ESR program. This conversion is done automatically by the system according to predefined mapping instructions.

In all programming phases, graphically supported simulation functions are available. They allow the verification of user inputs in the different programming phases. The whole programming task cannot be regarded as a first and final sequential execution of the single steps but rather as an iterative process with feedback.

Up to this point it is assumed that absolutely accurate geometrical data are available; tolerances of the robot itself and its environment were not taken into account. These ideal conditions are of course never really given. The conclusion is that off-line generated programs which have been tested by the simulation system cannot be executed in the real work cell. To solve this problem, calibration methods are used. The real model data are identified by the application of a measuring procedure. Using these data an error compensation of the application program as described in Chapter 15 is possible.

29.3   Application: Spot Welding of a Car Door

This application was used as the first testbed for off-line programming to realize the information flow from the CAD model of a workpiece down to the task execution of the real robot within a workcell.

The objective of this project was the generation of robot application programs for spot welding of a car door, their testing by the simulation system, their transfer to the robot controller and execution by an industrial robot. A welding cell was built containing a KUKA IR 160, an ARO C 1730 welding gun and the rear door of a Ford Sierra which was mounted on PTM module tables. The robot controller was a KUKA RC 20/41 which corresponds to the Siemens RCM 3.

Fig. 29.3   Realized test bed work cell.

By prior calibration of the robot, the off-line generated robot programs could be run without any additional on-line programming. Figure 29.3 shows the realized testbed workcell in reality and in simulation.

29.4   Application: Inspection of welding seams

This application was used to show the possibilities of transferring the off-line programming approach also to other programmable components, such as a submersible vehicle.

The overall objective of the so-called OSIRIS project is to make available an underwater working robot in connection with a qualified carrier /1/. The main tasks are the cleaning and inspection of welding seams at underwater structures without diver assistance. This requires the planning and programming of the submersible vehicle's motion as well as the task execution of the robot itself. Therefore the existing system was enlarged by the additional module 'vehicle programming' as shown in Figure 29.4.

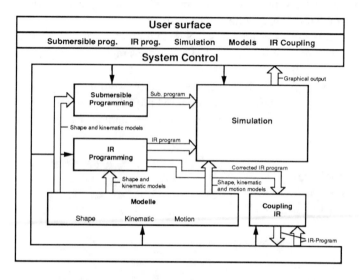

Fig. 29.4 Principle functional structure of the OSIRIS programming and simulation system.

As the first step of the programming procedure, the application program for the submersible vehicle is generated. The motion execution is tested via the simulation system (Fig. 29.5). In the next step, the same procedure is applied to generate and test the robot program. In Figure 29.6 a simulation of the modelled environment is shown. Additionally a 'work cell' was built in a laboratory consisting of a part of the industrial robot. Off-line generated and tested robot programs were transferred to the robot control and executed by the robot. Figure 29.7 shows a photo of the experimental set-up. The project was conducted by the partners Interatom, GKSS and IPK Berlin.

Fig. 29.5   Simulation of the submersible's motion.

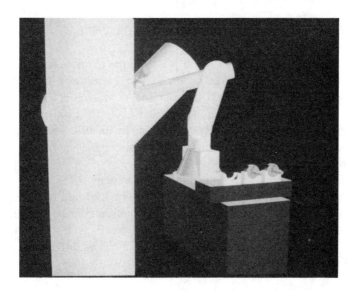

Fig. 29.6   Inspection robot with tool exchange system and underwater structure
(partly visible).

319

Fig. 29.7   Realized laboratory experiment.

29.5   Application: Assembly of a Matrix Printer

In the frame of the ESPRIT Project No. 623, the partners KUKA, PSI, IPK and the universities of Galway, Madrid and Lisbon realized an integrated planning and off-line programming system (Demonstrator A) which is described in Chapter 23. For an assembly task, the assembly of parts of a matrix printer was selected. The specific items of the project to be worked out by the IPK were as follows:
- planning modules for the selection of components and layout planning;
- simulation of the whole process and
- co-operation on the concept of the information system.

From the point of view of the simulation system, the main topic to be solved was the integration concept. The defined information interfaces, realized by a relational data base, consider the layout of the workcell, the task description, and the robot task program to be checked. Figure 29.8 shows different visualization outputs of the simulation module.

320

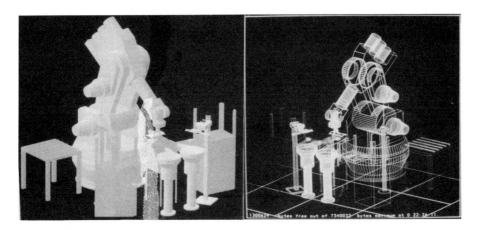

Fig. 29.8   Simulation of the assembly task execution.

29.6   Application: Robot Technology Experiment (ROTEX)

Robotics must be regarded as a key technology for automation in and commercial use of space in the future. Concerning production technologies, automation and teleoperation concepts have to be developed and tested. In this context the results achieved within the ESPRIT 623 project are helpful or can even be directly used for space applications.

For the German space-lab mission envisaged for 1992 a robot technology experiment is being developed, in which a robot in the experiment box (Fig. 29.9) has to fulfil different tasks after a calibration procedure. In Figure 29.10 details of the experiment cell are shown. The relevant robot programs have to be planned, generated and tested by simulation in the ground station and transferred for execution into orbit. The principal system structure is shown in Fig. 29.11. The project is financed by the German Ministry for Research and Technology with Dornier as the prime contractor and the IPK Berlin as a subcontractor of Dornier.

321

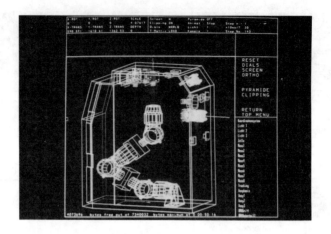

Fig. 29.9 ROTEX experiment simulation.

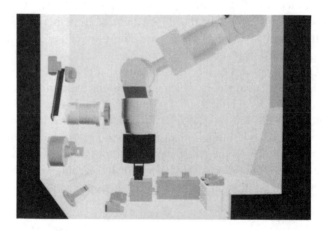

Fig. 29.10 Details of the ROTEX simulation.

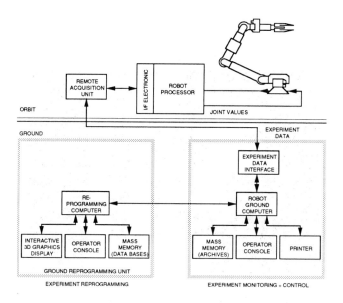

Fig. 29.11 System structure of ROTEX.

29.7   Application: Lab for in-ORbit Automation (LORA)

At the IPK Berlin a laboratory for automation and robotics (A&R) in space has been built which provides a development environment for automation procedures and components for space applications. This laboratory allows the development and testing of automation and experiments under realistic conditions. The project is partly supported by the Senate of Berlin, and a number of research institutions and Berlin SME's (Small and Medium Enterprises) participate in it. As a first experiment, a robotized work cell (5-axis-robot mounted on two external linear axes) for the exchange of samples from a melting stove has been realized. Thereby the task execution is planned and tested by the simulation system (Fig. 29.12). A photograph of the real work cell is shown in Fig. 29.13.

Fig. 29.12 Simulation of the task execution (LORA).

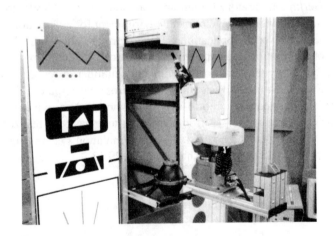

Fig. 29.13 Realized work cell (LORA).

29.8 Conclusion

As was shown, the off-line programming approach can be used in a wide area of applications. Inside ESPRIT 623, the concepts and basic modules of an off-line programming system have been realized. Furthermore, an infrastructure concerning skilled people in this area and high-performance soft- and hardware (especially in the graphics area) has been set up. The realized projects in lunar applications have to be seen as prototypes for the validation of specified reference models. The development of systems for a commercial end user require much more efforts and can only be realized by specific projects. Nevertheless, the work of ESPRIT 623 builds the basis for further developments in IPK in the area of off-line programming.

Chapter 30

Conclusions

R. Dillmann
University of Karlsruhe, Germany

The previous chapters described how an automatically operating software system or a human planner operating interactively with a computer can accomplish the planning of manufacturing tasks by referencing a generated world model. The strategy of planning is to decompose a manufacturing task successively into subtasks and to refine them further into basic motion commands. The strategy followed in this project is hierarchically structured. A set of modular tools is used to support the strategy. A working plan for an industrial robot application is to be decomposed into a sequence of smaller tasks to fulfil the desired functionality and to reach the goal. Each of these individual tasks is expanded and decomposed into a sequence of elementary actions to be performed by robots and peripherals. Elementary operations describe transfer, approach and departure movements as well as grasping, release, insert operations, welding operations or sensor-guided movements.

To support the activity of planning and progamming the robot task, various tools have been implemented and applied during the project. For different domains of applications the tools can be applied or adapted to the specific needs. The methods of successively planning and programming with the help of an information management system was successfully applied. The result is a generalized concept for integrated robot operational control.

The benefits of such an integrated system are shorter development costs for robot application development, lower risks and a better exploitation of robot technology. One example for a very successful planning and programming method is the use of interactive graphical simulation systems as shown in the previous applications.

With graphical robot simulation systems, the programmer has access to an interactive ergonomic tool to define and to modify program commands. The effects and correctness of these commands with respect to the desired robot task and the geometry of the product can be tested with the help of numerical and graphical simulation techniques.

Programming means specifying motion, interaction and the sequence of robot operations. Furthermore, a robot program has to determine the global sequence and the time requirements of the final robot program. An important aspect is that in this stage of program development various alternatives for layout, action sequences and motion can be considered and evaluated. This means that the final decision on the best alternative can be done without a high risk.

Further work in the project has been related to the use of AI tools to support decision making during planning and programming. Based on a generalized systems architecture, AI tools for layout planning, scheduling, action planning, geometrical reasoning for motion and interaction have been elaborated. From this platform new applications can be developed. A toolbox system consisting of the developed modules and integrated via an information system would be a consequent step forward. The role of neutral interfaces in such a toolbox system proposed by the STEP initiative would be helpful.

Further improvements of performance and efficiency can be expected by consequent object-oriented modelling methods and the use of advanced object-oriented tools. Data base management systems, CAD modellers, graphics, XPS shells, all implemented in C++, offer further improvements.

Section VI

Summary

Chapter 31

Major achievements

R. Bernhardt
IPK Berlin, Germany

The general goal of the project was to specify and to build prototype systems to demonstrate the integration of robots into CIM. At the beginning of the project, the three areas, <u>planning</u> and <u>programming</u> of a robotized system as well as <u>integration</u> into a CIM environment, were identified as the most important ones. The work focussed on building computer aided tools in the first two areas and to further develop them to an integrated planning and programming system. In all of the three areas two different approaches were adopted:

- The first was a more practical approach which made it very likely to achieve results which can be effectively used by the industry.
- The second approach was a more research oriented one, i.e. to use advanced methods like expert systems and AI technologies.Both approaches were characterized by a close cooperation of industrial and academic partners and in both practical results were achieved. For constructing the system, a two-way strategy was adopted. On the one hand, tools in the different areas had to be realized and on the other hand these tools had to be combined to show the increase in performance and efficiency of an integrated planning and programming system. Thereby it was an important requirement, specifically put forward by the industrial partners, not only to have a finished system at the end of the project but also to have intermediate results gained in the course of the project which were independently applicable on an industrial level. This feedback from practice had the positive effect of making refinements and improvements possible during the early development phases.

The first or bottom-up approach mentioned above concerns mainly the areas:

- interactive component selection and layout planning;
- explicit (motion oriented) off-line programming of industrial robots;
- simulation system to test the task execution in the robotized workcell, and
- realization of the integration vias a relational database management system.

But the system has been set up in such a way, and this has also partially been validated, that also sophisticated tools, e.g. for trajectory planning and optimization, can easily be integrated. These tools belong to the second or top-down approach which concerns mainly the areas:

- expert systems for assembly planning and component selection;
- implicit (task oriented) programming system, and
- knowledge based information system.

The realization of these advanced concepts was not also taken far enough to allow industrial application, but the interfaces of the tools have been designed in a way as to prepare them for integration.

The combination of both approaches led to a fruitful cooperation between industrial and academic partners for the following reasons: on the one hand results of industrial value could be reached which led to direct benefits for the everyday work in the companies and even to products available on the market. On the other hand, technical progress was achieved which did not take too much account of industrial constraints but which explored some areas in a more fundamental manner. This work, mainly performed by the academic partners, is of high interest for the industrial partners because the limitations in certain fields became better known and the decision making with regard to the actual application of those technologies was made easier.

Last but not least, a further achievement should be mentioned. For most of the partners, it was the first time that they cooperated in such a big international project. Furthermore industrial partners working in different industrial areas or even being competitors, and partners from different universities and research institutions having different areas of work and different backgrounds, speaking six different languages and having different cultural roots, organized a cooperation from scratch and in the end reached results valuable for both the industry and the academia. This was a very impressive experience, and also an encouraging one in view of future projects.

Chapter 32

Exploitation of results

R. Bernhardt
IPK Berlin, Germany

32.1   Introduction

According to a rule of the European Commission, the exploitation of results is divided into three categories. In the first one all those project results are gathered which directly contribute to a product or a service which is either already available on the market or going to be in the future. These results are described in section 32.2.

In the second category there are those results which are being used outside the team who developed them. Here, a further differentiation can be made between results used by another division of a company which was a member of the team and those results used by companies outside the project team. For these results, see section 32.3. In both categories the commercial value aspects are stressed, because results and experiences gained on the basis of which a partner can participate in national or international publicly funded projects are not regarded. However, these results are also very important especially for project partners from academia. In contrast to this, the use of results in industrial projects, i.e. a direct cooperation with another company, fall into the second category.

Lastly, results which contributed to the work of national or international standardisation boards belong to a third category. The activities undertaken in this area by the ESPRIT 623 project consortium are presented separately in Chapter 33.

Apart from these categories, project results made a very substantial contribution to the educational efforts of the academic partners involved in the project. Thirty-eight courses were either newly set up or re-arranged; 1700 students per year are taking part in these courses in the various European countries in which the academic partners reside. Additionally, more than 200 students participated directly in the project on a part-time contract basis. These activities were welcomed not only by the companies involved in the project but by the industry in general.

The exploitation results which fall into the aforementioned categories one and two are now presented in more detail.

32.2   Products and services

It is evident that from a commercial point of view project related products and services can only be provided by companies. Only they have the facilities to bring a product on the market and to ensure proper maintenance. These products are based on contributions to prototype systems by all project partners which were extended and customized by the industrial partners [1].

A product has been brought on the market by Renault Automation, Paris, which is called Robot Interactive Programming System. In a first step, this system was adapted to the requirements of the automotive industry. Following its successful application, it was then transferred to other industrial areas such as nuclear and aerospace industries. Furthermore the system has been applied in traditional industries which produce a small number of special parts with complex shapes. By means of the system, technologies such as laser cutting and high pressure water cutting can be effectively applied. The following example shows the benefits of the system: For a hardfacing application in which material has to be brought onto a complex surface at a temperature of 200°C, a robotized system was planned, programmed and built. Formerly, this work had to be done by welders and it took 6000 hours to do the job. With the robotized system this time was reduced to 2500 hours while achieving a better and constant quality. The planning of the welding trajectory performed by a welding engineer formerly took 16 hours. By using the planning functions of the system, this time was reduced to 2 hours. For further details, see Chapter 26.

KUKA at Augsburg, Germany, brought three products on the market as a result of the project The first, which is called KUSIM, is a system which supports the design, off-line programming and simulation of robotized work cells. As KUKA is very much involved in the planning and building of transfer lines for the automotive industry, KUSIM is specifically tailored for these applications but it can also be used in other areas. As KUKA is primarily a robot manufacturer, the company supplies also simulation models of industrial robots.

A further KUKA product is the so-called Robot Application Assistant (RAA). This PC based system is used for both altering existing and adapting off-line generated robot programs. The functions of this product allow the detection of variations in the real work cell by suitable sensors and perform a compensation by altering the program data. The system further features mirror imaging functions and the structural adaption (macro technique) of robot programs.

In Chapter 28, the benefits of the KUKA products are described in more detail. It may be highlighted here that the application of these tools for making a robotized work cell operational can result in time saving of up to 90%, depending on the manufacturing task. Accordingly, a substantial reduction of costs can be achieved. Furthermore, a decisive improvement of the quality of the solution can be reached as it is possible to analyze and compare alternative solutions. As these aspects are of great value to industry, project results and developed products were presented at 25 international fairs in a number of European countries. Applications and their benefits are presented in full detail in Section V.

32.3 Results achieved outside the project

In the course of the project a number of software tools and systems of different complexity were developed. Based on this software, more than 40 projects were carried out by the project partners concerning services (consulting, evaluation etc.) for other companies or institutions. Also in parallel to the ESPRIT 623 project, specific adaptations of software systems were made for direct cooperations with companies. In these cases the software was delivered to these companies together with the relevant system and user manuals. Additionally specific courses were held for employees from the recipient companies in order to ensure an efficient use of the systems and also further developments of the software. This process is still continuing even one year after the finishing of the project.

Until now, software tools and systems based on developments achieved in ESPRIT 623 have been delivered to fourteen European companies and five institutions in the educational and research domain. The applications are ranging from automotive, electrical/electronic and nuclear industries to shipbuilding and off-shore facilities and the aerospace industry [2, 3, 4, 5]. Some of these applications are described in more detail in Chapters 24 to 29.

References

[1] R. Bernhardt, *Manufacturing System Planning and Programming in a CIM Environment*, Proceedings of the Annual ESPRIT Conference, Brussels, November 12-15, 1990.

[2]   R. Bernhardt, *Integrated Planning and Off-line Programming System for Robotized Work Cells*, Proceedings of the Annual ESPRIT Conference, Brussels, November 27 - December 1, 1989.

[3]   B. Duffau, *Interactive Off-Line Robot Programming, Industrial Applications*, 5th CIM Europe Conference, Athens, Greece, 17-19 May 1989.

[4]   G. Duelen, R. Bernhardt, V. Katschinski, *Programming and Simulation System for the Automation and Robotics Technology Testbed*, Proceedings of the 1st Workshop on Robotics in Space, IARP, Pisa, Italy, June 1991.

[5]   G. Duelen, P. Armbrust, M. Imam, B. Loske, G. Schreck, H. Zander, *Off-Line Programmierung von Schweißrobotern im Schiffbau*, ZWF 86 (1991) 8, pp. 387-391.

Chapter 33

Standardization aspects

W. Jakob, B. Kottke
PSI Gesellschaft für Prozeßsteuerungs- und
Informationssysteme

In industry, robot programming is done mainly on-line by teach-in methods. This works well for more or less simple tasks such as pick and place or spot welding, but the resulting downtime causes high costs and becomes unacceptable for more complicated applications. This problem of high costs caused by downtime can be solved by off-line programming. Altough graphical off-line programming systems have been available for several years, these systems have not been really successful on the market. Knowing the obvious advantages of off-line programming, it is natural to ask why hardly anyone is using it.

33.1 Demands of the market

Based on a market study which was done by PSI in the second half of 1988 and general experience, it is possible to state:

1. There was no commonly accepted programming language for industrial robots or handling devices at the time. The absence of such a language is the reason for the gap between off-line and on-line programming; usually the programming interface on the off-line level differs from the on-line one. This is one of the most serious handicaps of off-line programming.
2. In particular, users who want to use robots from different vendors for their various applications stated their interest in a common robot programming language. Because the automobile industry counts in this group, they represent more than 50% of the market.

3. The CAD world differs from the world of real robots. This is not only a question of tolerances but also of ambiguities of tranformation algorithms for many commonly used robot kinematics. Usually, planning or programming tools solve them in a different way than their particular robot controller does.
4. Although there are some advantages in the area of standards for exchanging 3-D CAD data, there is still some work to be done, e.g. for exchanging kinematic data.
5. The development of an off-line programming system which is powerful enough to be of practical use is too expensive for a single robot control.

This is why standardised interfaces are the keypoint for the more general introduction of off-line programming; it must be possible to interface the off-line programming system to existing CAD systems as well as to robots without the need of two different languages. The introduction of standard interfaces will reduce the price for off-line programming tools and enhance their capabilities. This will make off-line programming a common technique which can also be used by medium- and small-size companies.

33.2   Role of Standard Interfaces in Robotics

The role of the three standard interfaces STEP, IRL and ICR is shown in Fig. 33.1. The kinematic information (K) and the geometric data describing the robot, tools, environment and move targets (G) are expressed by STEP. Motion oriented information (M) may be included.

Robot move instructors which describe not only the target but also a great variety of move parameters like speed or acceleration can be expressed by ICR and IRL. This holds also for technological parameters (T), program flow instructions or arithmetic or geometric calculations as well as for sensor data processing (C). The main difference between ICR and IRL is that IRL is a *programming language* and therefore is designed for human beings, and that IRL as an executable code can be implemented easier and cheaper on a robot control than IRL and of course can also be executed faster.

The interface to or between CAD systems is STEP, and geometric or kinematic information eventually together with some motion data, is transferred. The scope of STEP is very wide and the stated intent is to capture a computerised product model in a neutral form without any loss of completeness and integrity during the whole life cycle of a product.

A component for planning robot tasks may have the ICR or the IRL interface beside STEP. If ICR is used as an interface, the textual programming (via IRL) is done within this system, which means that some kind of textual programming and

338

compilation facilities are incorporated. In the case of the IRL interface the planning system generates motion instructions which a programmer may enhance by technological data. The advantages of using IRL as an interface of the planning component are:

- IRL can serve for *documenting purposes*;
- IRL is *readable*;
- IRL supports *modularization*.

The textual programming system enlarges the generated program by technological information and in particular adds the necessary program flow control instructions. If the target robot control is based on IRL, work is finished here and the program can be loaded and executed. Otherwise, it has to pass through the compiler which generates ICR out of IRL. Usually the interface to the robot control is ICR. The advantages are:

- ICR can be implemented more cheaply;
- ICR requires less amount of memory and lower speed processors;
- ICR can be executed faster than a high-level language;
- ICR was designed for a robot control structure of today as well as tomorrow.

If corrections to the user program are needed, the program has to be corrected at the IRL language level and compiled into ICR again.

For the use of standardized neutral interfaces, the software tools shown in Fig. 33.2 include routines for the generation and/or acceptance of information expressed by the standard formats. At present, the main stream of information runs from the user via IRL to the programming system and from there via ICR to the robot control or simulation.

33.3 Standardization activities

The history of CAD model transfer begins in the late 70s with the emphasis on design drawings. Strictly speaking, the information required to produce a design drawing and not the product information captured in the drawing was the original issue. The standard still most widely used is IGES. Two-dimensional and three-dimensional line geometry and to some extent also non-geometrical information can be transferred with IGES in most practical applications.

Another standard for transfer of similar information is SET. Transfer of surface information has also become feasible. It is now covered in IGES, SET and the VDA-FS standard. Based on the results of the ESPRIT 1 project CAD*I, the aim

Chapter 34

Future of robots in CIM

Dr. R. Bernhardt
IPK Berlin, Germany

The system robot consists of three subsystems: the mechanical arm including the drives and servo amplifiers, the robot controller, and the software tools for planning and programming of robot applications.

From a CIM-point of view, the first area seems to be of minor importance. But this is only partially true due to the share of cost of this subsystem in comparison with the overall system, which results in a limitation of the number of applications. Moreover, for off-line programming absolute pose accuracy has to be increased by approxiamely the factor 10. Attempts to achieve this by narrowing the tolerances in robot production would lead to a considerable increase in the price of the mechanical system. In contrast, in order to reduce the costs, either tolerances can be widened or the design of the mechanical system can be simplified (e.g. by reducing arm stiffness requirements). The goal therefore is a self-contradicting one: increase accuracy and reliability while reducing costs. Additionally, efficient quality control procedures in robot production have to be established. Reducing costs and increasing accuracy can only be achieved by adequate software tools for robot parameter identification and error compensation procedures in combination with enhanced features of the robot controller. Summarizing these thoughts, the mechanical system has to be improved by electronic and software means. Those software means themselves are part of the third area: the planning tools. They have to be integrated in a CIM environment as particularily since such tools are the true CIM components. Work in this area has already been started, for example in the the ESPRIT Projects CIM-SEARCH (5272) and CAR (5220).

Reducing costs by improving performance using by the means mentioned above will contribute tremendously to increasing the number of robot applications. This leads to a positive feed-back loop: an enlargement of the number of produced robots will automatically result in a system price reduction.

From a CIM point of view, the second subsystem, the robot controller as it is today, is perhaps the weakest component.

Robot controllers have to be very much improved concerning their features and functionalities. Thereby aspects like open system structure, hard- and software modularity, sensor integration and, last but not least, user support sytems for operation and maintenance play a dominant role. Improvements in these areas will also enlarge the number of robot applications, with the ensuing effects already mentioned. This requires, at least on a European level, a much closer cooperation between manufacturers of controllers and robots, production system builders and end users, and also a closer cooperation of the robot manufacturers among themselves.

The third area which is often characterized by the term 'off-line programming', is perhaps the most developed one. It covers the whole field of planning and operating a robotized cell, inclusive of the handing of exceptional cases. When talking about CIM and robotics, normally this area is addressed. During the last five to eight years many efforts have been undertaken all over the world. A large number of software tools and systems have been developed, many of which have already been industrially applied and have lead to remarkable benefits. Others, the more sophisticated ones, e.g. tools for trajectory planning and optimization, are still only available as laboratory versions. What is still missing, or at least is underdeveloped, is a stronger emphasis on the information integration aspects. These include on the information system side the provision of information exchange, safe data handling, data consistency etc., and furthermore aspects of uniform user interfaces. These areas can be characterized by the terms internal and external communication. For internal communication, work has been done in the past based on data base management systems which have already reached maturity for industrial application. Based on the gained experience, more advanced approaches - knowledge based information system - are in elaboration. Missing in this area is an information system model, or a paradigm, agreed upon by the researchers and industry. But efforts in this direction are already under way. For external communicaton, i.e. the communication with the planning engineer, the situation seems to be much easier, because standarized software packages and tools are already available on the market. This does not automatically mean that all problems have been solved, but the availability of such software supports the development process very much.

In both areas work is in progress in European R&D projects in the frame of the ESPRIT program, e.g. CIM-OSA, IMPPACT, CIDAM and CIM-PLATO.

The future of robots in CIM is strongly dependent on the enlargement of application fields, whereby the number of robots can be increased and their costs reduced. This requires improvements of all subsystems mentioned above. Such improvements can be reached by a close cooperation between research institutions and industry which has been proven to be fruitful since the start of R&D in robotics. Today one looks back on twenty years of robot development. If one compares this stretch of time with the time it took to develop automobiles, then today we would still be in a stage of development in which automobiles still closely resembled horse-carriages, in their form as well as in their function. It may be that the development of robots today is still in the phase of coaches. Nonetheless, the system robot has already proven its innovative force and its efficiency in many applications. If the rise in the number of robot applications worldwide since the early eighties is taken into account, one can draw the conclusion, that the application area of robots is steadily being enlarged. The developments of the recent years which are summarized under the acronym CIM include the implementation of computer-aided tools for planning, programming and simulation and the development of an integrated information system linking all the areas from design to the workshop, and the enlargement and improvement of the functionality and capacity to integrate the robot control. The discernable trend of development in the mentioned area will lead to new application areas for the system robot and probably marks the end of, as it were, the horse-carriage age of the robot. This means that, at least from a CIM point of view, the future of robots has just begun.

Index

346